Daniel Díaz Plascencia
Pablo Fidel Mancillas Flores
José Roberto Espinoza Prieto

Apple yeast in the diet of weaned calves

Daniel Díaz Plascencia
Pablo Fidel Mancillas Flores
José Roberto Espinoza Prieto

Apple yeast in the diet of weaned calves

Beneficial Yeasts in Beef Cattle Feed

ScienciaScripts

Imprint

Any brand names and product names mentioned in this book are subject to trademark, brand or patent protection and are trademarks or registered trademarks of their respective holders. The use of brand names, product names, common names, trade names, product descriptions etc. even without a particular marking in this work is in no way to be construed to mean that such names may be regarded as unrestricted in respect of trademark and brand protection legislation and could thus be used by anyone.

Cover image: www.ingimage.com

This book is a translation from the original published under ISBN 978-620-3-58859-0.

Publisher:
Sciencia Scripts
is a trademark of
Dodo Books Indian Ocean Ltd. and OmniScriptum S.R.L publishing group

120 High Road, East Finchley, London, N2 9ED, United Kingdom
Str. Armeneasca 28/1, office 1, Chisinau MD-2012, Republic of Moldova, Europe
Printed at: see last page
ISBN: 978-620-7-73017-9

Contents

GENERAL SUMMARY

This investigation consists of three experiments whose overall objective was to evaluate the inclusion of a yeast inoculum (YI) and fermented apple bagasse (BMZN) in the diet of early weaned calves and its effect on *in vitro* digestibility, performance and animal health. In the first experiment, the effect of an IL and BMZN in the diet of growing Angus calves on performance and immune system was evaluated. The behavioural test variables were analysed with a statistical model that included treatment as a fixed effect, treatment and sampling as fixed effects for the immune system variables, and animal as a random effect. The results showed that IL improved (P<0.05) daily feed intake (DFI) and feed conversion (FC). BMZN and IL increased (P<0.05) antioxidant activity (AA). On sampling day 84, T1 (oat hay (HA) + corn silage (MS) + concentrate and T2 (HA + MS + concentrate + BMZN) showed the highest (P<0.05) concentration of leucocytes (Leu) and lymphocytes (Lin). The second study consisted of evaluating the effect on *in vitro* fermentation of diets with the addition of an IL and BMZN. For the analysis of fibre digestibility variables, treatment was considered as a fixed effect, while for volatile fatty acids (VFA), N-NH3, lactic acid and pH, treatment and time were included as fixed effects. The results showed that the highest digestibility (P<0.05) of DM, NDF, FDA and lowest LDA content was presented by T2 and T3 (HA + DM + concentrate + IL). These showed higher (P<0.05) VFA concentration. The third experiment was to evaluate the effect on *in vitro* fermentation of growing calf diets supplemented with four yeast strains. The variables were analysed in a similar manner to the
second experiment. The highest DM digestibility (P<0.05) was presented by T2 (HA + EM + concentrate + *Kluyveromyces lactis* 2; Kl2), T3 (HA + EM + concentrate + *Klyveromyces lactis* 11; Kl11) and T4 (HA + EM + concentrate + *Issatchenkya orientalis* 3; Io3). The latter was superior (P<0.05) in NDF digestibility, furthermore, T2 and T3 increased FDA digestibility, therefore, the lowest (P<0.05) LDA content was presented by T2, T3 and T4. It is concluded that the inclusion of IL in the diet of growing calves improves CDA and CA, and together with BMZN increases AA. In addition, Kl2, Kl11 and Io3 improve the microbial environment favouring DM, NDF and FDA digestibility.

GENERAL INTRODUCTION

Food is one of the main needs of our population and in order to satisfy it, several domestic and wild animal species are exploited in the world, among which is the bovine cattle for its meat production. Cattle farming in Mexico represents one of the main activities of the agricultural sector, due to the contribution it makes to the supply of meat products, as well as its participation in live cattle exports (SIAP, 2012). At national level, an inventory of 23.3 million head of cattle was reported in 2008, 87.3 % dedicated to meat production and 12.7 % specialised in milk production, with the State of Chihuahua having 1.7 million head, of which 363,000 animals were meat producers (INEGI, 2009).

Beef cattle producers in Chihuahua are oriented towards the production of calves for export, where diets based on natural grasses and agricultural stools are typically used, limiting fibre digestibility and productive response of the animals (Dado and Allen, 1996). To maintain profitability on various types of livestock farms, producers in all sectors are looking for different options to reduce production costs (Gunn *et al.*, 2010). Fermented apple bagasse is an alternative as a protein supplement that can replace part of the ingredients in ruminant diets, with increases in milk production (Gutierrez, 2007) and animal health benefits (Gallegos, 2007). Added to this is the effect of yeast cultures which are rich in B vitamins, minerals and various types of amino acids.

(van der Peet-Schwering *et al.*, 2007) and as a consequence, stimulate nutrient absorption creating a healthy intestinal environment as well as a better immune system (Czarneki-Maulden, 2008).

Therefore, the objective of this study was to evaluate the effect of a yeast inoculum and fermented apple bagasse in the diet of Angus calves on: 1) productive behaviour and immune function, 2) *in vitro* digestibility of the three diets offered to Angus calves, 3) *in vitro* fibre digestibility using four different yeast strains.

Environmental Temperature on Productive Behaviour

Environmental temperature is one of the most researched variables and at the same time the most widely used as an indicator of stress. Mujibi *et al.* (2010) mentioned that the effects of environmental temperature on animal performance have been studied extensively in cattle, as ruminants within their genetic and physiological capacity continuously adjust to cope with environmental changes (Young *et al.*, 1989).

Arias *et al.* (2008) indicated that animals adapted to the environment in which they live, experience stress due to temperature oscillations or a combination of negative factors to which they are subjected for a short period of time, coping with it through physiological and behavioural modifications. Thus, in most cases the response reveals changes in nutrient requirements, with water and energy needs being most affected when cattle are outside the thermo-neutral zone (Conrad, 1985). Arias *et al.* (2008) demonstrated that changes in requirements, as well as the strategies adopted by the animals to cope with the stress period, result in a decrease in their productive performance. As part of the animal's acclimatisation behaviour, dry matter intake (DM) and daily water intake are directly affected, as both are related to heat balance and impact on body temperature regulation (Finch, 1986). Collin *et al.* (2001) found that in animals within their thermoneutrality zone, dietary energy is used for maintenance, growth, production and physical activity;

while below or above this zone energy is redirected to functions to maintain the homeothermic condition and in some cases there may be an increased energy demand for these processes.

Probiotics in Animal Feed

Diet plays an important role in animal and human health. In recent years, special attention has been devoted to the production of functional foods, aiming to add beneficial microorganisms or compounds into the body through daily dietary intake (Di Criscio *et al.*, 2010).

Probiotics have shown promising results in several areas of animal production. A probiotic is defined as a culture or mixed culture of live microorganisms that benefit humans or animals by improving the properties of the autochthonous gut microflora (Champagne *et al.*, 2005). In another study, it was mentioned that probiotics are viable microorganisms that are beneficial to the host when consumed in appropriate amounts, resulting in improved production and health (Rook and Burnet, 2005). Benefits include inhibition of pathogenic bacteria, reduction of serum cholesterol levels, diarrhoea and intestinal cancer; improved lactose tolerance, calcium absorption, vitamin synthesis and stimulation of the immune system (Sanchez *et al.*, 2009). Bontempo *et al.* (2006) suggested possible mechanisms that exhibit a beneficial role in the animal, indicating colonisation and adhesion in the intestinal mucosa (competition for receptors), competition for nutrients, production of antimicrobial substances and stimulation of the mucosal and immune system. Yeasts as probiotics are gaining popularity in fattening systems, as they are resistant and have a high viability under adverse environmental conditions. Comitini *et al.* (2005) reported that yeasts exert inhibitory activity on different strains of pathogenic bacteria, developing bactericidal or bacteriostatic activity due to certain metabolic compounds produced by them. Stephens *et al.* (2007) reported that lambs fed probiotics decreased the excretion of *Escherichia coli* O157:H7, the same authors published that direct microbial supplementation in feed also reduced the amount of *Salmonella ssp.* in beef cattle. The use of various probiotic species favours increased daily weight gain (DWG) and feed efficiency of lambs and fattening cattle during the initial feeding period (Lema *et al.*, 2001).

Apple Bagasse in Animal Supplementation

Manterola *et al.* (1999) described that apple bagasse is the residue from the apple juice extraction process and represents 15-20% of the processed fruit. Since long, there has been increasing interest in the use of apple bagasse in cattle diets; and like most feedstuffs, the usefulness of this by-product in ruminant diets depends on the fermentation processes in the rumen (Rumsey, 1978). Apple bagasse is equivalent to corn silage in total digestible nutrient content, deficient in digestible protein and higher in pectins, pentosans and ether extract than most common feeds (NAS, 1971). Sunvold *et al.* (1995) mentioned that there is little information on the fermentation of vegetable and fruit fibre. However, there is evidence that fruit and vegetable fibre contain several bioactive compounds (flavonoids and carotenoids) which increase their nutritional value in dog foods (Swanson *et al.*, 2001). Saura-Calixto and Larrauri (1996) demonstrated that fruit and vegetable fibre contain a balanced amount of soluble and insoluble fibre, which promotes gastrointestinal health. There is evidence that apple by-product can be added in diets of milk producing cows (12 to 15 L/d) up to 40 % of the dry matter of the radon (Edwards and Parker, 1995). Other researchers reported that supplementing milk producing cows with apple bagasse increased milk yields by 5.9 to 9 % and increased fat and protein content (Anrique and Dossow, 2003). Rodriguez *et al.* (2006) found that calves fed multinutritional blocks made with 14.6 % manzarine can have daily gains of 0.570 kg d. On the other hand, Gutierrez (2007) reported that fermented apple by-product can substitute part of the ingredients used in the feed of milk producing cows, having positive yields in production and animal health (Gallegos, 2007) due to its chemical and chemical-rich characteristics.

Yeast Cultures in Animal Nutrition

A wide variety of feed additives that manipulate rumen activity are available on the market. Ruminants establish a symbiosis with rumen microorganisms, whereby they provide nutrients and a suitable rumen environment for the survival of the microorganisms by improving feed fermentation, increasing the ability to utilise rumen-synthesised microbial fibre and protein, which is the main source of energy and protein (Calsamiglia *et al.*, 2005). Desnoyers *et al.* (2009) demonstrated in a compilation of 157 experiments that yeast supplementation increased feed intake, milk yield, rumen pH, rumen volatile fatty acids (VFA) and digestibility of organic matter; also, the effect of yeast was enhanced when animals consumed diets with a high proportion of concentrate. These same results suggested that yeasts limit the decrease in ruminal pH upwardly associated with an increase in VFA concentration and reduction of lactic acid, producing greater buffering capacity. Ingvartsen (2006) reported different nutritional strategies to help increase dry matter intake (DM) and avoid negative energy balance as much as possible during the first weeks of lactation in dairy cattle. However, many of these studies have been carried out on steers and dry cows fistulated or *in vitro* (Al Ibrahim *et al.*, 2010). These same authors mentioned that numerous commercial products are available, varying widely in *Saccharomyces cerevisiae* (Sc) strains, viability and number of cells present. Some Sc strains have favoured the establishment of cellulolytic bacteria in the digestive tract of lambs, accelerating microbial activity in the rumen, potentially favouring the switch from a liquid to a solid diet in preruminant animals (Chaucheyras-Durand and Fonty, 2001). On the other hand, Davis and Drackley (1998) mentioned that as the young calf matures and switches from a liquid diet to a grain and forage diet, the risk of diarrhoea tends to decrease.

Yeast Cultures in Rumen Fermentation

The inclusion of Sc in the feed results in an increase in the total number of cellulolytic bacteria (*Fibrobacter succinogenes, Ruminococcus albus*), both *in vitro* and *in vivo* (Lila *et al.*, 2004). In another study, Chaucheyras *et al.* (1995) observed the stimulation of the growth of the fungus *Neocallimastix frontalis*.

The same authors reported that Sc appears to stimulate lactate utilization by *Megasfaera elsdenii* and *Staphylococcus ruminantium* leading to an increase in propionate synthesis (Lila *et al.*, 2004). The latter authors reported that the decrease in lactic acid level results in an increase in ruminal pH favouring the growth of cellulolytic bacteria, leading to an increase in fibre digestion and in the production of volatile fatty acids (VFA). On the other hand, the effect of Sc on the concentration of ammoniacal nitrogen ($N-NH_3$) is very variable, and both a reduction and an increase have been observed (Chaucheyras-Durand and Fonty, 2001).

Calsamiglia *et al.* (2005) have reported evidence of the effect of yeasts on ruminal fermentation, mentioning that yeasts and their culture media may contain nutrients (organic acids, B vitamins, enzymes, amino acids, etc.) that stimulate the growth of bacteria that digest cellulose and utilise lactic acid. These same authors reported that live yeasts, through their respiration action, consume the residual oxygen available in the rumen medium, protecting the strict anaerobic bacteria. Newbold *et al.* (1996) compared several Sc strains and observed a strong correlation between the ability of yeasts to consume oxygen and bacterial growth, which allowed them to conclude that the stimulatory effect of Sc on ruminal bacteria could be attributed, at least partially, to its respiratory activity.

In another study, Yoon and Stern (1996) reported a mechanism of action for yeasts and fungi whereby increased rumen pH and reduced oxygen availability stimulate increased growth of cellulolytic bacteria, thereby improving fibre degradability, decreasing ruminal fullness, increasing dry matter intake and increasing production, without necessarily improving nutrient utilisation efficiency. Chung *et al.* (2011) published that yeast also has a potential to increase the fermentation process in the rumen in a way that decreases methane gas (CH_4) formation. In another study they proposed that, through strain selection, it may be possible to develop a commercial yeast product that decreases CH4 production while minimising rumen acidosis and promoting fibre digestion and rumen fermentation (Newbold and Rode, 2006).

Function of the Yeast Cell Wall in the Immune System

The yeast cell wall contains mannan-oligosaccharides (MOS) that can act as high-affinity receptors competing with binding sites for gram-negative bacteria, which possess mannose type 1 spiked fimbriae (Ofek *et al.*, 1977), eliminating pathogens from the digestive system and preventing colonisation and attachment of the pathogen to the mucosa (Nocek *et al.*, 2011). Ferket (2003) mentioned that this benefit can provoke an important antigenic response, thus enhancing humoral immunity against specific pathogens through the presence of antigens to attenuated immune cells, in addition, this process can suppress the proinflammatory immune response, which is detrimental to productive behaviour. Another predominant component of the yeast cell wall is e-1,3/1,6-glucan (β-glucan), which has been shown to have an immunomodulatory effect when yeast was used as a supplement in diets for poultry (Chae *et al.*, 2006), pigs (Li *et al.*, 2006) and aquatic animals (Dalmo and Bogwald, 2008).

Few studies have investigated the use of yeast cell wall components on immune function in dairy cattle (Nocek *et al.*, 2011). However, Franklin *et al.* (2005) observed that supplementation of dry cows with MOS improved humoral immune response against rotavirus and increased antibody transfer to their offspring. Reed and Nagodawithana (1991) published that oligosaccharides present in the cell wall of Sc such as glucans and mannans enhance the immune system and influence pathogen-host interaction in the digestive tract of animals and humans. Laboratory animals, which consumed oat glucans, enhanced neutrophil function and increased defence against pathogens (Murphy *et al.*, 2007). This may be particularly important in young calves, which are commonly affected by protozoa, viruses and bacteria that cause digestive tract diseases and some others that can lead to systemic infections (Magalhaes *et al.*, 2008). Likewise, soluble products from yeast cultures inhibit microbial activity and growth and modulate the immune system (Jensen *et al.*, 2008).

Function of Microminerals in the Immune System

Trace minerals (TM) such as copper (Cu), zinc (Zn), manganese (Mn) and cobalt (Co) are important for proper growth, reproduction and immune response (Dorton et al., 2007), and in many physiological processes in animals (Spolders, 2007). However, the impact of TM supplementation on ruminant immunity has been variable (Spears, 2000). This same author mentioned that trace element supplementation was initially focused on preventing signs of deficiency and decreased production, but the role of these elements in immunity has been emphasised in more recent studies.

The concentration of Zn and Cu in serum is influenced by many factors, such as the time of sampling and the animal (Spolders et al., 2010). On the other hand, blood components can be influenced by many external factors (climate, season and time of day) and internal factors such as breed, age and stage of lactation (McDowell, 2003).

MT deficiencies result in impaired growth rate (Blackmon et al., 1967), calving difficulties (James et al., 1987), and decreased production of T cells, B cells, neutrophils and macrophages, leading to increased susceptibility to infection (Chandra, 1999).

Zinc. It is an essential microelement for the proper functioning of cells, it has also been identified as a structural, catalytic or regulatory component of more than 200 enzymes involved in the metabolism of proteins, carbohydrates and nucleic acids (Vallee and Auld, 1990), being essential for the integrity of the immune system (Hambridge et al., 1986), although its specific role on the immune response is not very clear. Likewise, Murray et al. (2000) mentioned that it is a structural component of the enzyme superoxide dismutase (SOD), which helps to eliminate free radicals produced by various processes in the body. On the other hand, the protective function of Zn against the formation of free radicals and oxidative stress has led to studies on the antioxidant effect of Zn and its participation in the antioxidant defence system (de Oliveira et al., 2009). Considering the wide variety of Zn-containing enzymes (lactate dehydrogenase, alkaline phosphatase, alcohol dehydrogenase, carbon anhydrase, superoxide dismutase, carboxypeptidases, malic dehydrogenase, etc.), it is easy to imagine the serious consequences of cellular Zn deficiency (Kirchgessner et al., 1993).

Hambridge et al. (1986) published that Zn deficiency impairs immune function through a reduction in T cell function, as well as a decrease in the function of many key components of the immune system (thymus and neutrophils). Cymbaluk et al. (1986) mentioned that high Zn concentrations hinder Cu utilization and are accompanied by lameness, bone anaemia and can lead to anaemia (De Auer and Seawright, 1988).

Manganese. It is another potential Cu antagonist, therefore, ingredients with high Mn content in the diet may have a negative impact on Cu absorption (Hansen et al., 2009). Grace (1994) published that the Mn concentration of some forages can contain more than 100 mg/kg DM, while the Cu content is upwardly low. The antagonistic role of Mn on Cu is limited, however, studies with rats have shown a complicated interaction between the effect of dietary Mn (10 to 50 mg/kg DM) and Cu (<1 to 6 mg/kg DM) on indices of serum iron (Fe) levels (Reeves et al., 2004). Hidiroglow (1979) mentioned that Mn is poorly absorbed (1 % or less) in ruminant diets. In another study, they mentioned that dietary factors that may influence Mn bioavailability have received little attention, probably because deficiency is not considered a major problem in ruminants (van Bruwaene et al., 1984). On the other hand, limited evidence suggests that diets high in calcium and phosphorus may reduce Mn bioavailability (Hidiroglow, 1979).

Copper. It is an essential element in animal nutrition, and is involved in many biological processes in the body, mostly related to enzyme activities (Grace, 1994). Likewise, Swenson and Reece (1993) published that it is an integral part of the cytochrome system, Cu-superoxide dismutase (CuSOD), CuZn-superoxide dismutase (CuZnSOD), ceruloplasmin

(Cp), cytochrome oxidase, tyrosinase (polyphelin oxidase), ascorbic acid oxidase, plasma monoamine oxidase. Spears (2000) mentioned that in cattle and other species Cu has been documented to exert an effect on the immune system. In another study, it was found that the phagocytic ability of neutrophils doubled when Cu deficient calves were given Cu (Jones and Suttle, 1981). Prohaska and Failla (1993) reported that humoral immunity and mediator cells were dramatically decreased in Cu-deficient rats and mice. Also, the humoral immune response of growing steers was enhanced when injected with 90 mg Cu before weaning and fed 7.5 mg Cu/kg DM during the growth phase (Ward and Spears, 1999).

Cu deficiency in cattle is a fairly common disorder worldwide and their diets are regularly high in Cu concentration (above 35 mg/kg DM), the maximum level of Cu supplementation for cattle being set by the European Union (Council Regulation (EC) No. 1334/2003/EC), well above the general physiological requirements (10 mg/kg DM; NRC, 1996). Kendall *et al.* (2001) commented that the relatively wide margin in Cu supplementation is because in cattle, and in ruminants in general, nutritional requirements do not depend exclusively on the concentration of Cu in the diet, but are highly dependent on Cu supplementation and availability of Cu, which justifies interference with antagonists, mainly molybdenum (Mo), sulphur (S), iron (Fe) and Zn. Hansen *et al.* (2009) published that Cu deficiency in beef cattle can present itself in the form of reduced growth, hair loss and anaemia.

This sign can be explained by the reduced activity of cuproenzymes such as cytochrome c oxidase, tyrosinase and ceruloplasmin, which are important in energy production, melanin and Fe metabolism, respectively (NRC, 1996). **Antioxidant Activity**

Nutrition has a major effect on health and immunity in animals, nutritional deficiencies impair immune response and therefore increase morbidity and mortality (Chew, 1995). The same author mentioned that antioxidants serve to stabilise highly reactive free radicals, thereby maintaining the functional and structural integrity of cells. Antioxidants have been grouped into enzymatic and non-enzymatic, and are substances capable of controlling the production of free radicals (FR) in the organism that are generated as a consequence of aerobic metabolism, either by sequestering FR or stabilising them (Halliwell and Whiteman, 2004). Tremellen (2008) mentioned that to counteract the RLs there is in the first instance the enzymatic antioxidant system, which includes the seleno-enzyme glutathione peroxidase (GPx) that acts primarily by reducing hydrogen peroxide (H_2O_2), and SOD that acts on the superoxide anion (O_2) transforming it into a secondary radical (H_2O_2) for subsequent action of GPx.

Most of the molecules present in the organism are chemically stable, that is, they have an even number of electrons in their external orbital, but there are others whose number of electrons is odd; these highly unstable and reactive chemical substances are known as free radicals (Barquinero, 1992). Halliwell (1992) reported that when a radical interacts with a stable substance, it takes an electron from it, positively charging the molecule; thus, the stable compound becomes a highly aggressive free radical, which can generate more radicals, giving rise to a chain reaction. On the other hand, oxygen, although essential for life, is also toxic because it is an oxidising substance, as it can accept electrons, destabilising the molecule that loses them. Therefore, in aerobic metabolism, oxidants called reactive oxygenated metabolites are produced, among which are O_2, hydroxide (OH) and H_2O_2 (Ceballos *et al.*, 1998).

Confinement and heat stress can contribute to increased antioxidant requirements in animals. Therefore, studies have been conducted in which antioxidant supplements are offered to animals, either in the diet or parenterally, and it has been observed that supplementation improves the immune response and decreases oxidative stress, leading to increased resistance to infectious and degenerative diseases (Chew, 1995). Likewise, Gonzalez-San Jose *et al.* (2001) published that as an individual ages, the balance is shifted in favour of oxidants, making it of interest to ingest foods with antioxidants to decrease them. In another study, they mentioned that vitamin E is a soluble phenolic Kpid antioxidant; therefore its antioxidant activity is based on the ability to donate a hydrogen atom to free radicals (Yurttas *et al.*, 2000). **Haematic Biometry in Ruminants**

In the clinical practice of veterinary medicine, the parameters of haemometric biometry (Hb) values are a fundamental diagnostic aid in the analysis and orientation of the clinical status of an animal and in the monitoring of a herd; this information is useful in cases of control, evaluation of enzyme stability processes and the nutritional status of animals. BH is also a diagnostic aid in the evaluation of the herd, which allows prophylactic and curative decisions to be made. Barrio *et al.* (2003) mentioned that BH is the numerical and descriptive evaluation of the cellular elements of the blood (red blood cells, white blood cells and platelets); constituting one of the most requested tests in the clinical laboratory, since it accompanies almost all diagnostic protocols with precision, accuracy and speed.

Blood cell components transport oxygen (red blood cells or erythrocytes), protect against foreign organisms and viruses (white blood cells or leukocytes), phagocytose, capture or destroy them, and initiate coagulation (Aiello and Mays, 2000). These same authors mentioned that blood cells are divided into leukocytes (phagocytes and lymphocytes); phagocytes are subdivided into monocytes (mononuclear) and granulocytes (polymorphonuclear), and the latter are subdivided into neutrophils, eosinophils and basophils. On the other hand, lymphocytes are white blood cells responsible for humoral and cellular immunity, their production originating in the bone marrow. In another study, Aiello and Mays (2000) reported that haematological analyses show the quantities of cellular elements; their evaluation allows to determine health problems, tissue inflammation or proliferative functions of the bone marrow.

Rumen Digestibility and Rumen Fermentation of Foodstuffs

For some years, ruminant microbiologists and nutritionists have been interested in manipulating the rumen microbial ecosystem to improve production efficiency (Callaway and Martin, 1997). The addition of *Aspegillus oryzae* extracts and Sc cultures in ruminant diets improved DM digestibility, crude protein (CP) and hemicellulose; increasing the number of bacteria in the rumen, decreasing lactate concentration and increasing milk production in fresh cows (Gomez-Alarcon *et al.*, 1990). However, the response to supplementation with yeast or fungal cultures has been variable (Martin and Nisbet, 1992). Previous studies (Newbold *et al.*, 1996) have shown that yeast cultures increase the number of cellulolytic bacteria in the rumen, and in some cases increase cellulose degradation. On the other hand, digesta-attached microorganisms represent more than 75 % of the total rumen microflora, and Sc stimulates the growth of main y *F. succinogenes, R. albus and R. flavefaciens* being the most active in fibre degradation (Chaucheyras- Duran and Fonty, 2001). Also, other authors have used Sc yeast as an additive in fibrous diets for ruminants producing improvements in the efficiency of utilisation and availability of nutrients (Newbold *et al.*, 1998) and increases in rumen digestion of DM, organic matter and NDF, both in *vivo* and *in vitro*, as well as in the degradability of FDA and nitrogen (Biricik and Turkman, 2001).

Gas Production and VFA Profiles in the Rumen

Quantification of gas production in *in vitro* fermentations is known to be used to determine digestibility and rumen fermentation kinetics of feeds (Theodorou *et al.* 1994), which together with multiple regression equations and nutritional components of substrate (Noguera *et al.*, 2004) allow quite accurate estimation of *in vivo* degradability and apparent digestibility of DM of forages (Blummel *et al.*, 1997); fitting the accumulated gas profiles to an appropriate equation to summarise the kinetic information (Groot *et al.*, 1996).

On the other hand, energy for microbial growth is derived from the fermentation of carbohydrates, mainly starch and cellulose, whose anaerobic digestion produces VFA, succinate, lactate, ethanol, carbon dioxide (CO_2), CH4 and trace hydrogen (H_2); however, these also provide carbon skeletons essential for microbial biomass synthesis (Opatpatanakit *et al.,* 1994). In another study (Getachew *et al.,* 1998) reported that gas production is mainly generated when the substrate is fermented to acetate and butyrate, with propionate being produced only from the neutralisation of the acid and is therefore of lower quantity. On the other hand, ruminal microorganisms are very sensitive to changes in pH and most prefer a range between 6.5 and 6.8. Grant and Mertens (1992) commented that cellulorhizal bacteria, in particular, are more sensitive to low pH than amylopholytic bacteria. Likewise, Hoover *et al.* (1984) showed that high (7.5) or low (5.5) pH severely compromises fibre digestion.

Noguera *et al.* (2004) reported that the gas production technique allows the detection of differences between substrates generated by maturity, growth conditions, species or crop and preservation methods. It has also been used to determine differences in the fermentation of crop residues subjected to various chemical or physical treatments (Williams, 2000).

LITERATURE CITED

Aiello, S. E. and A. Mays. 2000. The Merck Veterinary Manual. 5th ed. Oceano Grupo Editorial, S. A. Spain.

Al Ibrahim, R. M., A. K. Kelly, L. O'Grady, V. P. Gath, C. McCamey and F. J. Mulligan. 2010. The effect of body condition score at calving and supplementation with *Saccharomyces cerevisiae* on milk production, metabolic status, and rumen fermentation of dairy cows in early lactation. J. Dairy Sci. 93:5318-5328.

Anrique, G. R. and C. C. Dossow. 2003. Effect of apple pulp silage in the ration of dairy cows on intake, replacement rate and milk production. Arch. Med. Vet. 35:13-22.

Arias, R. A., T. L. Mader and P. C. Escobar. 2008. Climatic factors affecting the productive performance of beef and dairy cattle. Arch. Med. Vet. 40:7-22.

Barquinero, J. 1992. Free Radicals: A Threat to Health. In: Crystal, R. G. and J. R. Ramon. 1992. GSH System. Glutathione: Axis of Antioxidant Defence. Excerpta Medica. Medical Communications B. V., Amsterdam. Kingdom of the Netherlands.

Barrio, M., M. C. Correa and M. E. Jimenez. 2003. The haemogram: analysis and interpretation of blood values. University of Antioquia, Medellin. Colombia.

Biricik, H. and I. I. Turkmen. 2001. The effect of *Saccharomyces cerevisiae* on *in vitro* rumen digestibilities of dry matter, organic matter and neutral detergent fiber of different forage: concentrate ratios in diets. Veteriner Fakultesi Dergisi, Uludag University. 20:29-37.

Blackmon, D. M., W. J. Miller and J. D. Morton. 1967. Zinc deficiency in ruminants, occurrence, effects, diagnosis, and treatments. Vet. Med. 62:265-272.

Blummel, M., H. P. S. Makkar and K. Becker. 1997. *In vitro* gas production: A technique revisited. J. Anim. Physiol. Anim. Nutr. 77:24-34.

Bontempo, V., A. Giancamillo, G. Savoini, V. Dell'Orto and C. Domeneghini. Domeneghini. 2006. Live yeast dietary supplementation acts upon intestinal morpho-functional aspect and growth in weaning piglet. Anim. Feed Sci. Technol. 129:224236.

Callaway, E. S. and S. A. Martin, 1997. Effects of a *Saccharomyces cerevisiae* culture on ruminal bacteria that utilize lactate and digest cellulose. J. Dairy Sci. 80:2035-2044.

Calsamiglia, S., L. Castillejos and M. Busquet. 2005. Nutritional strategies to modify ruminal fermentation in dairy cattle. Page 161 in Memoria XXI del curso de especializacion FEDNA. Barcelona. Barcelona. Spain.

Ceballos, A., F. Wittwer, P. A. Contreras and T. M. Bohmwald. 1998. Blood glutathione peroxidase activity in grazing dairy herds: variation according to season and time of year. Arch. Med. Vet. 30:13-22.

Chae, B. J., J. D. Lohakare, W. K. Moon, S. L. Lee, Y. H. Park and T. W. Hahn. 2006. Effects of supplementation of beta-glucan on the growth performance and immunity in broilers. Res. Vet. Sci. 80:291-298.

Champagne, C. P., N. J. Gardner and D. Roy. 2005. Challenges in the addition of probiotic cultures to foods. Crit. Rev. Food Sci. Nutr. 45:61-84.

Chandra, R. K. 1999. Nutrition and immunology: From the clinic to cellular biology and back again. Proc. Nutr. Soc. 58:681-683.

Chaucheyras, F., G. Fonty, G. Bertin and P. Gouet. 1995. *In vitro* H2 utilization by a ruminal acetogenic bacterium cultivated alone or in association with an *Archaea methanogen* is stimulated by a probiotic strain. Appl. Environ. Microbiol. 61:3466-3467.

Chaucheyras-Durand, F. and G. Fonty. 2001. Establishment of cellulolytic bacteria and development of fermentative activities in the rumen of gnotobiotically- reared lambs receiving the microbial additive *Saccharomyces cerevisiae* CNCM I-1077. Reprod. Nutr. Dev. 41:57-68.

Chew, B. P. 1995. Antioxidant vitamins affect food animal immunity and health. J. Nutr. 125:1804-1808.

Chung, Y. H., N. D. Walker, S. M. McGinn and K. A. Beauchemin. 2011. Differing effects of 2 active dried yeast (*Saccharomyces cerevisiae*) strains on ruminal acidosis and methane production in nonlactating dairy cows. J. Dairy Sci. 94:2431-2439.

Collin, A., J. Van Milgen, S. Dubois and J. Noblet. 2001. Effect of high temperature on feeding behaviour and heat production in group-housed young pigs. Br. J. Nutr. 86:63-70.

Comitini, F., R. Ferretti, F. Clementi, I. Mannazzu and M. Ciani. 2005. Interaction between *Saccharomyces cerevisiae* and malonic bacteria: preliminary characterization of a yeast proteinaceous compounds active against *Oenococcus oeni*. J. Appl. Microbiol. 99:105-111

Conrad, J. H. 1985. Feeding of farm animals in hot and cold environments. In: Yoursef MK 5th ed. Stress Physiology in Livestock Volume II Ungulates. CRC Press Boca Raton, Florida, USA.

Council Regulation (EC) No 1334/2003/EC on amending the conditions for authorization of a number of additives in feedingstuffs belonging to the group of trace elements. Off. J. Eur. Union.

Cymbaluk, N. F., F. M. Bristol and D. A. Christensen. 1986. Influence of age and breed of equid on plasma cooper and zinc concentrations. Am. J. Vet. Res. 47:192-195.

Czarnecki-Maulden, G. L. 2008. Effect of dietary modulation of the intestinal microbiota on reproduction and early growth. Theriogenology. 70:286290.

Dado, R. G. and Allen M. S. 1996. Enhanced intake and production of cow offered ensiled alfalfa with higher neutral detergent fiber digestibility. J. Dairy Sci. 79:418-428.

Dalmo, R. A. and J. Bogwald. 2008. Beta-glucans as conductors of immune symphonies. Fish Shellfish Immunol. 25:384-396.

Davis, C. L. and J. K. Drackley. 1998. The Development, Nutrition and Management of the Young Calf. Iowa State Press. Ames. Iowa. U.S.A.

De Auer, J. C. and A. A. Seawright. 1988. Assessment of copper and zinc status of farm horses and training thorouhbreds in south-east Queensland. Aust. Vet. J. 65:317-320.

de Oliveira K. J., C. M. Donangelo, A. V. de Oliveira, Jr., C. L. de Silveira, and J. C. Koury. 2009. Effect of zinc supplementation on the antioxidant, copper, and iron status of physically active adolescents. Cell Biochem. Funct. 27:162-166.

Desnoyers, M., S. Giger-Reverdin, G. Bertin, C. Duvaux-Ponter and D. Sauvant. 2009. Meta-analysis of the influence of *Saccharomyces cerevisiae* supplementation on ruminal parameters and milk production of ruminants. J. Dairy Sci. 92:1620-1632.

Di Criscio, T., A. Fratianni, R. Mignogna, L. Cinquanta, R. Coppola, E. Sorrentino and G. Panfili. 2010. Production of functional probiotic, prebiotic and synbiotic ice creams. J. Dairy Sci. 93:4555-4564.

Dorton, K. L., T. E. Engle, R. M. Enns and J. J. Wagner. 2007. Effects of trace mineral supplementation, source, and growth implants on immune response of growing and finishing feedlot steers. The Professional Animal Scientist. 23:29-35.

Edwards, J. and W. Parker. 1995. Apple pomace as a supplement to pasture for dairy cows in late lactation. Proc. N. Z. Soc. Anim. Prod. 55:67-69.

Ferket, P. R. 2003. Controlling gut Health with the Use of Antibiotics. Pag. 57-68 in Proc. 30th Annu. Carolina Poult. Nutr. Conf., Research Triangle Park, NC. North Carolina State University, Raleigh. U.S.A.

Finch, V. A. 1986. Body temperature in beef cattle: its control and relevance to production in the tropics. J. Anim. Sci. 62:531-542.

Franklin, S. T., M. C. Newman, K. E. Newman, and K. I. Meek. 2005. Immune parameters of dry cows fed mannan oligosaccharide and subsequent transfer of immunity to calves. J. Dairy Sci. 88:766-775.

Gallegos, A. M. A. 2007. Somatic cell count in milk, plasma antioxidant activity and blood cellular components of Holstein cows in production fed manzarin in the diet. Master's thesis.

Faculty of Animal Science. Autonomous University of Chihuahua. Chihuahua. Chihihuahua, Chih. Mex.

Getachew, G., M. Blummel, H. P. Makkar, and K. Becker. Becker. 1998. *In vitro* measuring techniques for assessment of nutritional quality of feeds: a review. Anim. Feed Sci. Technol. 72:261-281.

Gomez-Alarcon, R. A., C. Dudas and J. T. Huber. 1990. Influence of cultures of *Aspergillus oryzae* on rumen and total tract digestibility of dietary components. J. Dairy Sci. 73:703-710.

Gonzalez-San Jose, M. L., R. P. Muniz and V. B. Valls. 2001. Antioxidant activity of beer: *in vitro* and *in vivo* studies. Beer and Health Information Centre. Department of Biotechnology and Food Science. University of Burgos. Department of Paediatrics, Gynaecology and Obstetrics. University of Valencia. Spain.

Grace, N. 1994. Managing Trace Element Deficiencies. New Zealand: AgResearch, New Zealand Pastoral Agriculture Research Institute Ltd. New Zealand.

Grant, R. J. and D. R. Mertens. 1992. Development of buffer systems for pH control and evaluation of pH effects on fiber digestion *in vitro*. J. Dairy Sci. 75:1581-1587.

Groot, J. C., J. W. Cone, B. A. Williams, F. M. Debersaques, and E. A. Lantinga. 1996. Multiphasic analysis of gas production kinetics for *in vitro* fermentation of ruminant feeds. Anim. Feed Sci. Technol. 64:77-89.

Gunn, P. J., M. K. Neary, R. P. Lemenager, and S. L. Lake. 2010. Effects of crude glycerin on performance and carcass characteristics of finishing wether lambs. J. Anim. Sci. 88:1771-1776.

Gutierrez, P. F. J. 2007. Effect of manzarin on physicochemical components and milk production. Master Thesis. Faculty of Animal Husbandry. Autonomous University of Chihuahua. Chihuahua, Chih. Mex.

Halliwell, B. 1992. Reactive Oxygen Species in Living Things: Origin, Biochemical and Pathogenic Role in Man. In: Cristal, R. G. and J. R. Ramon, 1992. GSH System. Glutathione: Axis of Antioxidant Defence. Excerpta Medica. Medical Communications B. V., Amsterdam. Kingdom of the Netherlands.

Halliwell, B. and M. Whiteman. 2004. Measuring reactive species and oxidative damage *in vivo* and cell culture: how should you do it and what do the results mean? Br. J. Pharmacol. 142:231-255.

Hambridge, K. M., C. E. Casey and N. F. Krebs. 1986. Zinc. In trace elements in human and animal nutrition. 5th ed. Walter Mertz. Academic press, Inc., London. U. K.

Hansen, S. L., M. S. Ashwell, L. R. Legleiter, R. S. Fry, K. E. Lloyd and J. W. Spears. 2009. The addition of high manganese to a copper-deficient diet further depresses copper status and growth of cattle. British J. Nutr. 101:10681078.

Hidiroglow, M. 1979. Manganese in ruminant nutrition. Can. J. Anim. Sci. 59:217-236.

Hoover, W. H. H., C. R. Kincaid, G. A Vargas, W. H. Thayne and L. L. Junkins. 1984. Effects of solids and liquid flows on fermentation in continuous culture. IV. pH and dilution rate. J. Anim. Sci. 58:692-699.

INEGI. 2009. United Mexican States, Agricultural Census. VIII Agricultural, Livestock and Forestry Census. Aguascalientes, Ags., Mexico.

Ingvartsen, K. L. 2006. Feeding and management related diseases in the transition cow: Physiological adaptations around calving and strategies to reduce feeding related diseases. Anim. Feed Sci. Technol. 126:175-213.

James, S. J., M. Swendseid and T. Makinodan. 1987. Macrophage-mediated depression of T-cell proliferation in zinc-deficient mice. J. Nutr. 117:19821988.

Jensen, G. S., K. M. Patterson and I. Yoon. 2008. Nutritional yeast culture has specific anti-microbial properties without affecting healthy flora. Preliminary results. Anim. Feed Sci. Technol. 17:247-252.

Jones, D. G. and N. F. Suttle. 1981. Some effects of copper deficiency on leukocyte function in sheep and cattle. Res. Vet. Sci. 31:151-156.

Kendall, N. R., D. V. Illingworth and S. B. Telfer. 2001. Copper responsive infertility in british cattle: the use of a blood caeruloplasmin to copper ratio in determining a requirement for copper supplementation. In: Diskin, M. G. (ed), Fertility in the high-producing dairy cow. British Society of Animal Science, vol. 26(2). Occasional Publication, Edinburgh. United Kingdom.

Kirchgessner, M. P. and B. R. P. Roth. R. P. Roth. 1993. Zinc in animal nutrition. Sci. Invest. Agr. 20:182-201.

Lema, M., L. Williams and D. R. Rao. 2001. Reduction of fecal shedding of enterohemorrhagic *Escherichia coli O157:H7* in lambs by feeding microbial feed supplement. Small Rumin. Res. 39:31-39.

Li, J., D. F. Li, J. J. Xing, Z. B. Cheng and C. H. Lai. 2006. Effects of beta-glucan extracted from *Saccharomyces cerevisiae* on growth performance, and immunological and somatotropic responses of pigs challenged with *Escherichia coli* lipopolysaccharide. J. Anim. Sci. 84:2374-2381.

Lila, Z. A., N. Mohammed, T. Yasui, Y. Kurokawa, S. Kanda and H. Itabashi. 2004. Effects of a twin strain of *Saccharomyces cerevisiae* live cells on mixed ruminal microorganism fermentation *in vitro*. J. Anim. Sci. 82:1847-1854.

Magalhaes, V. J. J. A., F. Susca, F. S. Lima, A. F. Branco, I. Yoon and J. E. P. Santos. 2008. Effect of feeding yeast culture on performance, health, and immunocompetence of dairy calves. J. Dairy Sci. 91:1497-1509.

Malecki, E. A. and J. L. Greger. 1996. Manganese protects against heart mitochondrial lipid peroxidation in rats fed high levels of polyunsaturated fatty acids. J. Nutr. 126:27-33.

Manterola, B. H., A. D. Cerda and J. J. Mira. 1999. Agricultural Residues and their Use in Ruminant Feeding. Apple Pomace (*Malus pumila*), Use in Milk Producing Cattle. Faculty of Agronomical Sciences. University of Chile. Chile.

Martin, S. A. and D. J. Nisbet. 1992. Effect of direct-fed microbials on rumen microbial fermentation. J. Dairy Sci. 75:1736-1744.

McDowell, L. R. 2003. Minerals in Animal and Human Nutrition, 2nd ed. Elsevier Science B. V., Amsterdam, The Netherlands. V., Amsterdam, The Netherlands. Kingdom of the Netherlands.

McNaught, C. E. and J. MacFie. 2001. Probiotics in clinical practices: a critical review of evidence. Nutr. Research. 21:343-353.

Morelli, L. 2002. Probiotics: Clinics and/or nutrition. Dig. Liver Dis. 34:8-11.

Mujibi, F. D. N., S. S. Moore, D. J. Nkrumah, Z. Wang and J. A. Basarab. 2010. Season of testing and its effect on feed intake and efficiency on growing beef cattle. J. Anim. Sci. 88:3789-3799.

Murphy, E. A., J. M. Davis, A. S. Brown, M. D. Carmichael, A. Ghaffar and E. P. Mayer. P. Mayer. 2007. Oat в-glucan effects on neutrophil respiratory burst activity following exercise. Med. Sci. Sports Exerc. 39:639-644.

Murray, R. K., D. K. Granner, P. A. Mayers, and V. W. Rodwell. 2000. W. Rodwell. 2000. Harper's Biochemistry. 25th ed. McGraw Hill Health Professional Division, New York, USA.

NAS, 1971. Atlas of Nutrition Data on United States and Canadian feeds. National Academy of Sciences, Washington, DC. U.S.A.

Newbold, C. J., F. M. McIntosh and R. J. Wallace. 1998. Changes in the microbial population of a rumen-simulating fermenter in response to yeast culture. Canadian J. Animal Sci. 78:241-244.

Newbold, C. J., R. J. Wallace and F. M. Mcintosh. 1996. Mode of action of the yeast *Saccharomyces cerevisiae* as a feed additive for ruminants. British J. Nutr. 76:249-261.

Newbold, C. J. and L. M. Rode. 2006. Dietary Additives to Control Methanogenesis in the Rumen. Pages 138-147 in Greenhouse Gases and Animal Agriculture: An update. Elsevier, Amsterdam. The Netherlands. Kingdom of the Netherlands.

Nocek, J. E., M. G. Holt and J. Oppy. 2011. Effects of supplementation with yeast culture and enzymatically hydrolyzed yeast on performance of early lactation dairy cattle. J. Dairy Sci. 94:4046-4056.

NRC. 1996. Nutrient Requirements of Beef Cattle, 7th ed. Washington, DC: National academy Press. U.S.A.

Ofek, I., D. Mirelman and N. Sharon. 1977. Adherence of *Escherichia coli* to human mucosal cells mediated by mannose receptors. Nature. 265:623- 625.

Opatpatanakit, Y., R. C. Kellaway, I. J. Lean, G. Annison and A. Kirby. 1994. Microbial fermentation of cereal grains *in vitro*. Aust. J. Agric. Res. 45:1247-1263.

Noguera, R. R., E. O. Saliba and R. M. Mauricio. 2004. Comparison of mathematical models for estimating degradation parameters obtained through the gas production technique. Livestock Research for Rural Development. Vol. 16, Art. No. 86. http://www.lrrd.org /lrrd16/11/nogu16086.htm. Accessed October 10, 2012.

Prohaska, J. R. and M. L. Failla. 1993. Copper and Immunity. In: Klurfeld, D. M. Human Nutrition a Comprehensive Treatise, 5th ed. Plenum press, New York. U. S. A.

Reed, G. and T. Nagodawithana. 1991. Yeast Technology. 2nd ed. AVI, Van Nostrand Reinhold Publ. New York. U.S.A.

Reeves, P. G., N. V. C. Ralston, J. P. Idso and H. Lukaski. 2004. Contrasting and cooperative effects of copper and iron deficiencies in male rats fed different concentrations of manganese and different sources of sulfur amino acids in an AIN-93G-based diet. J. Nutr. 134:416-425.

Rodriguez, M. C., J. F. Lucero, A. N. Melendez, H. E. Rodriguez, C. G. Hernandez and O. B. Ruiz. 2006. Forage intake and weight gain in commercial export calves supplemented with multi-nutrient blocks made with manzarin. Memorias, page 86, XXXIV annual meeting of the mexican association of animal production and X biennial meeting of the north mexican group of animal nutrition. Universidad Autonoma de Sinaloa, Mazatlan, Sinaloa, Mexico.

Rook, G. A. and L. R. Burnet. 2005. Microbes, immunoregulation and the gut. Gut. 54:317-320.

Rumsey, T. S. 1978. Ruminal fermentation products and plasma ammonia of fistulated steers fed apple pomace-urea diets. J. Anim. Sci. 47:967-976.

Sanchez, B., C. G. De Los Reyes-Gavilan, A. Margolles and M. Gueimonde. 2009. Probiotic fermented milks: Present and future. Int. J. Dairy Technol. 62:472-483.

Saura-Calixto, Y. and J. A. Larrauri. 1996. New types of high quality dietary fibres. Rev. Alim. Equip. Technol. 1:71-74.

SIAP. 2012. Livestock Population. Cattle, 1999-2008. http://www.siap.gob.mx/index.php?opt on=com_content&view=article&id = 3&Itemid=29. Accessed September 15, 2012.

Spears, J. W. 2000. Micronutrients and immune function in cattle. Proc. Nutr. Soc. 59:587-594.

Spolders, M. 2007. New Results of Trace Element Research in Cattle. In: M. Furll (ed.), Proceedings of 13th International Conference on Production Diseases in Farm Animals, Leipzig, Germany.

Spolders, M. S. Ohlschlager, J. Rehage and G. Flachowsky. 2010. Inter- and intra-individual differences in serum copper and zinc concentrations after feeding different amounts of copper and zinc over two lactations. J. Anim. Physiol. Anim. Nutr. 94:162-173.

Stephens, T. P., G. H. Longeragan, E. Karunasena and M. M. Brashears. 2007. Reduction of *Escherichia coli O157* and *Salmonella* in feces and on hides of feedlot cattle using various doses of a direct-fed microbial. J. Food Prot. 70:2386-2391.

Sunvold, G. D., G. C. Fahey, Jr., N. R. Merchen, and G. A. Reinhart. 1995. *In vitro* fermentation of selected fibrous substrates by dog and cat fecal inoculum: Influence of diet composition on substrate organic matter disappearance and short-chain fatty acid production. J. Anim. Sci. 73:1110-1122.

Swanson, K. S., C. M. Grieshop, G. M. Clapper, R. G. Shields, Jr., T. Belay, N. R. Merchen and G. C. Fahey, Jr. 2001. Fruit and vegetable fiber fermentation by gut microflora from canines. J. Anim. Sci. 79:919-926.

Swenson, M. J. and W. O. Reece. 1993. Duke's Physiology of Domestic Animals. 11th ed. Comstoch Publications Assoc. U.S.A.

Theodorou, M. K., B. A. Williams, M. S. Dhanoa, A. D. B. McAllan and J. France. 1994. A simple gas production method using a pressure transducer to determine the fermentation kinetics of ruminant feeds. Anim. Feeds Sci. Technol. 48:185-197.

Tremellen, K. 2008. Oxidative stress and male infertility-A clinical perspective. Hum. Reprod. Update. 14:243-258.

Vallee, B. L. and D. S. Auld. 1990. Zinc coordination, function and structure of zinc enzymes and other proteins. Biochem. 29:5647-5659.

van Bruwaene, R., G. B. Gerber, R. Kirchmann, J. Colard and J. van Kerkom. 1984. Metabolism of Cr. Mn. Fe and Co in lactating dairy cows. Health Phys. 46:1069-1082.

van der Peet-Schwering, C. M. C., A. J. M. Jansman, H. Smidt and I. Yoon. 2007. Effects of yeast culture on performance, gut integrity, and blood cell composition of weanling pigs. J. Anim. Sci. 85:3099-3109.

Waller, K. P., U. Gronlund and A. Johannisson. 2003. Intramammary infusion of beta 1,3-glucan for prevention and treatment of *Staphylococcus aureus* mastitis. J. Vet. Med. B. 50:121-127.

Ward, J. D. and J. W. Spears. 1999. The effects of low-copper diets with or without supplemental molybdenum on specific immune responses of stressed cattle. J. Anim. Sci. 77:230-237.

Williams, B. A. 2000. Cumulative Gas Production Techniques for Forage Evaluation. In: Givens D. I., Owen E, Omed H. M. and Axford R. F. E. (ed). Forage Evaluation in Ruminant Nutrition. Wallingford. Wallingford. United Kingdom.

Y oon, I. K. and M. D. Stern. 1996. Effects of *Saccharomyces cerevisiae* and *Aspergillus oryzae* cultures on ruminal fermentation in dairy cows. J. Dairy Sci. 79:411-417.

Y oung, B. A., B. Walker, A. E. Dixon and V. A. Walker. 1989. Physiological adaptation to the environment. J. Anim. Sci. 67:2426-2432.

Y urttas, H. C., H. W. Schafer and J. J. Warthesen. 2000. Antioxidant activity of nontocopherol hazelnut *(Corylus spp.)* phenolics. J. Food Sci. 65:276280.

**YEAST INOCULUM AND FERMENTED APPLE BAGASSE IN THE
DIET OF GROWING ANGUS CALVES ON
PRODUCTIVE PERFORMANCE AND IMMUNE SYSTEM**

SUMMARY

The objective of the study was to evaluate the effect of a yeast inoculum (YI) and fermented apple bagasse (BMZN) in the diet of growing Angus calves on performance and immune system. Twenty-six four-month-old calves (BW=112±6.2 kg) were randomly distributed into three treatments: T1 (control, n=9), T2 (BMZN, n=9) and T3 (IL, n=8), fed the following diets: T1: oat hay (HA) + corn silage (EM) + concentrate; T2: HA + EM + concentrate + BMZN; T3: HA + EM + concentrate + IL. The diets were isoproteic (28.8 % CP) and isoenergetic (1.26 Mcal/kg DM). The variables evaluated were live weight (BW), daily weight gain (DWG), daily feed intake (DFA), feed conversion (FC) and serum Zn, Mn and Cu concentrations. In addition, antioxidant activity (AA) in blood plasma and haematic biometry (BH) in whole blood were quantified. The variables were evaluated every 28 d of the test, except BH (56 and 84 d). Behavioural test variables were analysed with a statistical model that included treatment as a fixed effect. In the concentration of Zn, Mn, Cu, AA and BH, treatment and sampling were included as fixed effects, and the animal as random effect. T3 was superior (P<0.05) in CDA and CA (8.512±0.12 kg/d and 1.530±0.03, respectively). In Zn, T1, T2 and T3 showed the lowest concentrations (P<0.05) on the 56th day of sampling (11.14±2.77, 10.56±2.81 and 12.30±2.90 pmol/L, respectively). Mn showed a significant difference (P<0.05) on the 28th sampling day, with T2 and T3 showing the lowest concentration (4.29±2.11 and 7.28±2.22 pmol/L). On sampling day 28, T1, T2 and T3 showed lower (P<0.05) Cu concentrations (8.04±2.34, 8.32±2.34 and 9.76±2.47 pmol/L). The AA revealed treatment effect (P<0.05) on day 84 of the test, with T2 and T3 being higher with 15.72±0.03 and 15.71±0.03 mmol/L, respectively. T1 and T2 showed the highest concentration (P<0.05) of Leu on d^a 84 (9.82±0.95 and 11.11±0.95 x 10^3 /pL respectively) and Lin (7.30±0.92 and 8.73±0.92 x 10^3 /pL). It is concluded that T3 increased CDA and CA, on the other hand, maintained the amount of Leu and Lin close to the maximum level, without showing significant variations over time.

INTRODUCTION

Probiotics are feed additives containing microorganisms that when added to animal feed remain active in the digestive tract, exerting important physiological effects that contribute to the balance of the rumen environment and the immune system, with beneficial effects on production performance (Desnoyers *et al.*, 2009); however, the mechanism of the effect of yeast products is not fully understood. Live cultures of *Saccharomyces cerevisiae* produce enzymes, B vitamins, minerals and various types of amino acids (van der Peet-Schwering *et al.*, 2007) and as a consequence, stimulate nutrient absorption creating a healthy intestinal environment and enhancing the immune system (Czarnecki-Maulden, 2008). On the other hand, trace minerals such as Zinc, Copper and Manganese are elements that play an important role in various body functions necessary to maintain optimal health and therefore are essential nutrients for all animals (Aksu *et al.*, 2011), exerting their effect on proper growth, reproduction, hormone secretion pathways and immune response (Dorton *et al.*, 2007). Similarly, antioxidants may be part of enzymes that directly or indirectly protect cells against the adverse effects of numerous drugs and carcinogenic substances (Huerta *et al.*, 2005). The addition of antioxidants to the diet of domestic animals improves immune response and decreases oxidative stress, leading to increased resistance to infectious and degenerative diseases (Chew, 1995). Fermented apple bagasse is rich in antioxidants, organic minerals and yeasts that can help maintain the health of the animal that consumes it. It is low cost and has nutrients that are highly fermentable by microorganisms such as yeasts and bacteria in the rumen (Becerra *et al.*, 2008 and Rodriguez-Muela *et al.*, 2010).

Based on this background, the objective of this study was to evaluate the effect of a yeast and fermented apple bagasse inoculum in the diet of growing Angus calves and its effect on performance, serum mineral concentration, plasma antioxidant activity and haematic biometry. The results of this research will allow producers and livestock nutritionists to make a decision to improve the performance and immune system of calves subjected to weaning stress.

MATERIALS AND METHODS

Location of the Study Area

The present study was developed in the facilities of the ranch La Canada, located in the municipality of Guerrero, Chihuahua, Chih, Mexico. Located in the coordinates 28° 33' 05" north latitude and 106° 30' 07" west longitude, with an altitude of 2,010 masl, according to global positioning system (GPS, MAGELLAN®, MobileMapper-Pro).

The climate is semi-humid temperate; the average annual temperature is 13 °C, with an average maximum of 42 °C and an average minimum of -17.6 °C. The average annual precipitation is 517.2 mm, with a relative humidity of 65 % and an annual average of 90 days of rain (INAFED, 2010). The experiment started on 17 February and ended on 25 May 2011.

Description of Animals and Treatments

Twenty-seven weaned Angus calves with initial weight of 112±6.2 kg and average age of four months were used. Calves were randomly assigned to three treatments; T1 (control, n=9): oat hay (HA) + corn silage (EM) + concentrate; T2 (BMZN, n=9): HA + EM + concentrate + BMZN and T3 (IL, n=8): HA + EM + concentrate + IL. Concentrates (Table 1) were added to HA and EM at the time of offering the feed.

The BMZN was prepared with 75 kg of waste apple, to which 350 g of urea + 100 g of ammonium sulphate + 200 g of premix of vitamins and trace minerals were added, then mixed in an aerobic fermenter for 72 h, maintaining the mixing for 15 min at each time (0, 6, 12, 24, 48 and 48 h), and then mixed for 15 min at each time (0, 6, 12, 24, 48 and 48 h).

Table 1. Composition of concentrate feed by treatment

Ingredients	T1	T2	T3
% (BS)			

Ma^z rolled	32.0	30.6	32.0
Molasses	10.0	5.0	8.0
Flouroline 41 % p.c.	20.6	15.0	20.6
Soybean meal 44 % w/w	30.0	30.0	30.0
Calcium carbonate	2.4	2.4	2.4
BMZN[2]	0.0	12.0	0.0
Common salt	2.0	2.0	2.0
Minerals and vitamins12:10[1]	3.0	3.0	3.0
Yeast inoculum	0.0	0.0	2.0
Calculated analysis			
ENg Mcal/kg	1.26	1.25	1.26
PC % P.C.	28.81	28.83	28.89
PD % PD % PD % PD % PD %			
PD % PD % PD % PD % PD %	53.91	51.78	53.07
PD % PD %			
Ca % Ca % Ca % Ca % Ca %			
Ca % Ca % Ca % Ca % Ca %	1.67	1.77	1.68
Ca % Ca			
P %	0.98	0.99	0.99
Cu mg/kg	16.12	13.75	15.32
Mn mg/kg	31.66	30.18	31.17
Zn mg/kg	42.48	40.32	42.70

[1]Vitamin and mineral mix 12:10: P, Ca and Mg (12.0, 11.5, 0.6 %), Mn, Zn, Fe, Cu, I, Co and Se (2160, 2850, 580, 1100, 102, 13, 9 ppm), Vitamin A, D3 and E (220000, 24500, 30 IU/kg).
[2]BMZN = fermented apple bagasse.

72 h). This quantity of mixture is prepared in two stages, the first to prepare the concentrate for the months of February and March (600 kg), and the second for the months of April and May (600 kg).

Biological Material

The four yeast strains used for this work were obtained from the solid-state fermentation of apple bagasse (BM), which were identified through the extraction and amplification of 18S rDNA by polymerase chain reaction (PCR) in the animal transgenesis laboratory of the Facultad de Zootecnia y Ecolog^a of the Universidad Autonoma de Chihuahua. For identification, a yeast culture was performed by serial dilution and 16 colonies were isolated from which DNA was extracted to amplify a region of the 18S rDNA (752 bp). The PCR product obtained was subjected to sequencing and analysed using the Blast programme of the *National Center for Biotechnology Information* (NCBI) database.

Sequence analysis showed that the yeasts corresponding to the 16 isolated colonies were: *Saccharomyces cerevisiae; Issatchenkia orientalis* and *Kluyveromyces lactis* (Villagran *et al.*, 2009). The strains used for this work were *K. lactis* strains 2 and 11; *I. orientalis* strain 3 and *S. cerevisiae* strain 6, all obtained from solid-state fermentation of WB. The strains were maintained viable by periodic reseeding in cradles and petri dishes. The medium used was Malt extract at 33.6 g/L and the incubation time was 48 hours at 30 °C. The strains were then stored at 30 °C. The strains were incubated for 48 hours. Subsequently, they were kept refrigerated at a temperature of 4 °C. The IL used in the present experiment was prepared in 5 jars of 20 L, to 4 of them were added yeast cultures from 3 Petri boxes each of the strains described above, adding to each of the 5 jars, 6 g of urea, 1 g of ammonium sulphate, 500 g of cane molasses, 2.5 g of premix of vitamins and minerals and 6 L of distilled water. Each pot was fitted with a portable oxygenator for 96 h in order to supply the oxygen necessary to promote yeast growth.

Subsequently, 10 mL samples were taken from each jar in hermetically sealed 50 mL plastic bottles, which were transferred to the Animal Nutrition Laboratory to count yeasts by microscopy, with the help of a micropipette (Nichiryo LE) with a volume of 10 to 100 pL and disposable tips, A serial dilution of 1 mL of sample was prepared using distilled water as diluent, then 10 pL of the dilution were taken from each sample and placed in a hematortmeter (Neubauer chamber) for counting (Diaz, 2006).

The number of individual or pooled cells were counted in the 4 quadrants (1 mm^2 each) as indicated in Figure 1. The average per quadrant was determined and the number of cells per millilitre (cell/mL) was calculated, adjusting the number of each strain to 3.7 x 10^9 cells/mL in 500 mL by dilutions with the yeast-free substrate, to add 2 L of inoculum to the T3 concentrate.

Animal Management

Prior to the start of the experiment, calves were weighed and identified with plastic ear tags, vaccinated intramuscularly (IBR, BVD, PI3, VRSB) with a dose of 2 ML/animal, dewormed internally and externally via subcutaneous with Ivermectin 1 % (1 mL/50 kg of BW). Subsequently, the calves were randomly divided into three groups of nine calves, each group consisting of three pens (18 m^2) with a dirt floor,

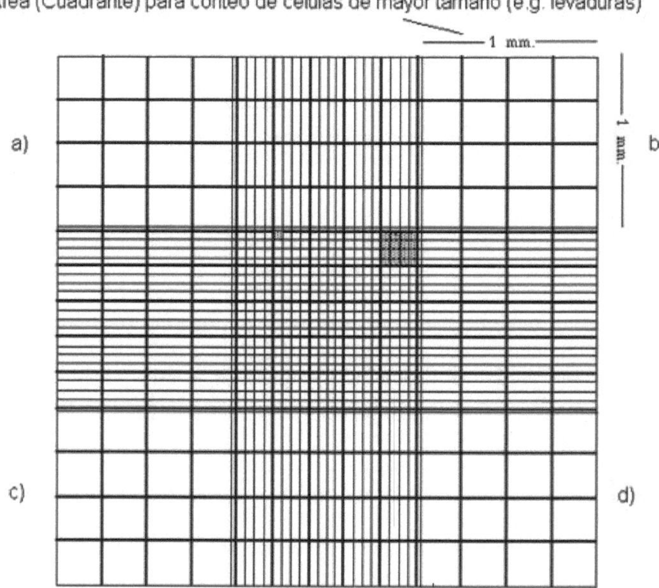

Figure 1. Grid of a haematocytometer (Neubauer chamber). a, b, c and d, were the four quadrants used for yeast counting, their surface area totalling 4 mm2, the space between a coverslip and the surface of the Neubauer chamber, using the boundaries of each quadrant as a guide is 0.1 mm3 in volume.

The animals were fed *ad libitum* once a day at 9:00 h. Three growth diets were used during the experiment (NRC, 1996), offering 1.2 kg HA and 1.7 kg ME on a dry matter (DM) basis, with the addition of 1.2 kg DM and 1.7 kg ME on a dry matter (DM) basis. Three growth diets were used during the experiment (NRC, 1996), offering 1.2 kg HA and 1.7 kg ME on a dry matter (DM) basis, adding 1.2 kg DM concentrate per animal per day according to the corresponding treatment. Every 14 d, feed intake was adjusted by weighing the feed offered and rejected for three consecutive days, where only the kg of HA and SM were increased by

15%, ensuring diet availability throughout the day.

One calf was dropped from the trial, due to causes unrelated to the treatment effect, so the sample size of T3 was 8 calves.

Sampling

Ambient temperature. The maximum and minimum ambient temperature (RT) was recorded daily with a mercury thermometer at 7:00 am and 1:00 pm.

Productive behaviour. Behavioural test variables were measured at baseline and every 28 d throughout the experiment. Calves were weighed individually on days 1, 28, 56 and 84 of the test, using a REVUELTA® scale with a 1500 kg capacity to determine live weight (BW). Feed offered and rejected were recorded three consecutive days per pen at the middle of each weighing period using a PEXA hand scale to calculate dry matter intake (DMI), adjusting for 10% rejection.

To calculate daily weight gain (DWG), a subtraction was made between consecutive weights and the product was divided by 28, being the number of days that elapsed between weighings. The information obtained on CMS and GDP was used to determine the feed conversion ratio (FCR), which is the ratio between the amount of feed consumed and the weight gain over a period of 28 days.

Blood samples. Four tubes of blood samples were collected per animal by direct puncture of the jugular vein after individual weighing and before they received the diet during the morning, these samples were preserved in a cooler stocked with ice.

Determination of minerals in serum. For the measurement of Zinc (Zn), Manganese (Mn) and Copper (Cu), two Vacutainer tubes® (without anticoagulant) were collected at the beginning and every 28 d of the test. The samples obtained were centrifuged at 3500 xg for 10 min at 4 °C to extract the blood serum; the supernatant was collected in pre-labelled 10 mL amber vials and subsequently frozen at - 20 °C until analysis. Prior to their measurement, they were thawed in refrigeration at 4 °C, then three subsamples of 2 mL for each mineral (Zn, Mn and Cu) were taken in 5 mL tubes (BD Falcon), adding trichloroacetic acid (ATCA) at 20 % (1 mL ATCA: 1 mL of serum), vortex mixing (Vortex Genie II) for 10 seconds (s) and then centrifugation at 3500 xg for 1C min and 4 °C to obtain a deproteinised supernatant, then the supernatant was diluted with distilled water in the same ratio (1 mL:1 mL).The analysis was performed with standard prepared glycerol to maintain the viscosity characteristics of the diluted samples. All solutions used were prepared with tridistilled water. The standard solution of Zn ($ZnSO^{\wedge}7H2O$) was prepared with 1.09 g diluted in 250 mL of tridistilled water, with the addition of 5 % glycerol; the standard of Mn ($MnSO^{\wedge}O$) was prepared with 0.77 g diluted in 250 mL of tridistilled water, adding 10 % glycerol; and the standard solution of Cu ($CuSO4^{\wedge}O$) was prepared with 9.71 mL diluted in 100 mL of distilled water, adding 10 % glycerol. The working solutions of the standards were prepared one day before analysing the samples. All samples were analysed in duplicate by atomic absorption spectrophotometry (AAnalyst 200, Perkin-Elmer instruments), following the recommendations of Makino and Takahara (1981) with results reported in parts per million (ppm).

Measurement of plasma antioxidant activity (AA): For the measurement of antioxidant activity (AA), a sample was collected in a Vacutainer tube® (7.2 mg EDTA as anticoagulant) at the beginning and every 28 d of the experiment. The samples destined to determine AA were transferred to the animal nutrition laboratory of the Faculty of Zootechnics and Ecology where they were centrifuged at 3500 xg for 10 min at 4 °C, the plasma was decanted in amber vials of 10 mL capacity, previously labelled, and then frozen at - 20 °C until the moment of its measurement. This variable was analysed with the technique developed by Benzie and Strain (1996), which aims to measure the ability of plasma to reduce iron or FRAP (Ferric Reductive Ability of Plasma). This technique is performed by colorimetry, diluting the plasma samples in distilled water and methanol (73.75 % distilled water, 25 % methanol and 1.25 % plasma).

Solution buffer (300 mM C H_{23} NaO2^H_2 O, pH 3.6), 3.1 g of sodium acetate trihydrate was added, 16 mL of glacial acetic acid (C2H4O2) was added and made up to 1 litre (L) with distilled water, this solution was preserved at 4 °C; 2.- Solution of 40 mM hydrochloric acid (HCl), 1.46 mL of concentrated HCl was added to a 1 L flask and made up to 1 litre (L) with distilled water, this solution was preserved at room temperature; 3.- Solution tris-pyridyl-triazine (TPTZ; 10 mM/L of 2,4,6- trispyridyl-s-triazine), 0.031 g of TPTZ was mixed in 10 mL of solution 2, this reagent was prepared at the time of use; 4.- Solution of ferric chloride hexahydrate (20 mM FeC^6H2O), 0.054 g were mixed in 10 mL of distilled water, this solution was prepared one day before use and kept refrigerated (4 °C).

The FRAP solution was composed of solutions 1, 3 and 4 in a ratio of 10:1:1 (v/v/v/v), this reagent was prepared at the time the samples were prepared for analysis, as it cannot be preserved.

The standard solution for the calibration curve was prepared with 0.0417 g of ferrous sulphate heptahydrate (FeSO<7H_2 O) diluted in 50 mL of HPLC grade methanol (CH3OH), with this mixture samples of 0.0 (blank), 0.2, 0.4, 0.6, 0.8 and 1.0 mL, to these quantities methanol was added (except to the last one) until completing 1 mL and later to each one 3 mL of distilled water was added, they were shaken in vortex for 10 s, later each point of the curve was analysed in triplicate, extracting 200 pL of each point and adding 1,800 pL of FRAP solution was added to each tube in triplicate, again vortexed and after 10 min of rest at room temperature, the reading was taken in a Junior® II Coleman spectrophotometer model 6/20 at an absorbance of 593 nanometres (nm). With the data obtained, a prediction curve was elaborated based on a regression equation.

Once the plasma was thawed, the samples were prepared, 7 pL of plasma were taken and added to a tube in which 200 pL of water had been previously added, then 195 pL of methanol were added and 195 pL of methanol were mixed and vortexed, 2,000 pL of FRAP solution were added to this mixture, after 10 min of rest the reading was taken at an absorbance of 593 nm. The blank was prepared in the same way as the samples, except that 7 pL of distilled water was added, the samples and the blank were prepared in triplicate. Using the equation and the amount of absorbance of the sample, the AA per micromolar (pm) of blood plasma was calculated, the results were expressed in micromolar Fe2 equivalents (pm Fe2).

Haematic biometry (Hb). To analyse haematic biometry (BH) a Vacutainer® tube (7.2 mg EDTA as anticoagulant) was collected on days 56 and 84 of the experiment. Different blood components were determined in quantity, percentage and weight as follows: Leukocytes (Leu), Neutrophil (Neu), Lymphocytes (Lin), Mixed cells (Cm), Erythrocytes (Er), Haemoglobin (Haem), Haematocrit (Hto), Mean corpuscular volume (MCV), Mean corpuscular haemoglobin (MCH), Mean corpuscular haemoglobin concentration (MCHC) and Platelets (Pq). The BH samples were sent to ASSAY Laboratorio Cloco S. de R. L. de C. V., where they were analysed using a Beckman Coulter cytometer® /act/dif/1al. **Statistical Analysis**

The variables of the behavioural test, such as PV, CDA, GDP and CA were analysed taking the treatment as a fixed effect and the initial weight as a covariate in a completely randomised design. On the other hand, Zn, Mn, Cu, AA and BH were analysed taking as fixed effects treatment and sampling, and their interaction, and as random effect the animal, Tukey's test was used to establish the possible differences between the means of the treatments (Steel and Torrie, 1997).

RESULTS AND DISCUSSION

Ambient Temperature

The average AT recorded during the months of this experiment is shown in Table 2. AT is probably the most researched variable and at the same time the most widely used as an indicator of stress. The effects of ambient temperature on animal performance have been studied extensively in cattle (Mujibi *et al.*, 2010), and it has been found that an animal, within its genetic and physiological capacity, continuously adjusts to cope with environmental changes (Young *et al.*, 1989). On the other hand, Arias (2006) mentioned that the effective ambient comfort temperature is the constant state of body temperature, which can be maintained without the need for physiological or behavioural adjustments.

In this study it is likely that the animals experienced cold stress, and in response increased feed intake to generate metabolic heat, which is consistent with Young (1983) who reported that animals exposed to low temperatures trigger various thermoregulatory mechanisms whereby maintenance requirements remain unchanged until the critical temperature is exceeded; conversely, tissue damage, immune system damage and decreased reproductive and growth response may occur (Moberg, 1987).

Behavioural Testing

There was a significant effect (P<0.05) on ADC and AC as a consequence of diet (Table 3). T3 calves were superior to T2 calves.

Table 2. Experimental   average environmental   registered   during the temperature

Date	Sampling $_{dayob -}$	Temperature °C[a]	
		Maxima	**Minima**
17-February / 01-March	1	24.3	-7.3
02-March / 30-March	28	27.2	-3.0
31-March / 27-April	56	30.4	0.9
28-April / 25-May	84	30.7	2.6
Average		28.1-1.7	

[a] The maximum and minimum temperature corresponds to the average recorded over the course of the days of the dates indicated.

bSampling days for the duration of the experiment.

Table 3. Means (± SE) of productive performance of calves fed three different diets.

Variables	Treatments[II]		
	T1	T2	T3
Live weight, kg	184.96±4.34[a]	178.34±4.36a	183.60±4.6ia
ADC, kg/d	7.89±0.12[b]	7.54±0.12[b]	8.51±0.12a
GDP, kg/d	0.86±0.05a	0.79±0.05a	0.85±0.05a
CA	1.39±0.03[b]	1.38±0.03[b]	1.53±0.03a

(P<0.05) in ADC (8.512 ±0.12 kg/d) and BW (1.530±0.03) than those offered fermented apple bagasse and the control diet. However, although calves that received an IL in their diet showed an effect on ADC and CA, there was no (P>0.05) difference in BW and GDP when compared to the other two treatments. It is possible that the IL stimulated the numerical growth of cellulolytic bacteria and increased fibre fermentation at the rumen level, thereby increasing feed intake without improving BW and BW.

According to other research, the increase in daily feed intake observed is the result of an increase in ruminal degradation of fibre due to the addition of the yeast culture (Newbold *et al.*, 1996), which seems to be associated with the stimulation of growth and activity of cellulolytic bacteria, increasing the flow of protein into the small intestine, which is expected

---

[a,b] Means with different row literals are different (P<0.05) between treatments.

[II] T1 = Oat hay (HA) + corn silage (EM) + concentrate; T2 = HA + EM + concentrate + fermented apple bagasse (BMZN); T3 = HA + EM + concentrate + yeast inoculum (IL).

to improve the animal's development due to the increased availability of nutrients (Bm^k and Urkmen, 2001). Young (1983) indicated that animals exposed to cold conditions below body temperature are associated with metabolic acclimatisation, which translates into increased heat production to maintain the body's vital functions. Delfino and Mathison (1991) reported evidence that low temperatures lead to poor feeding efficiency behaviour.

In the present study, the extreme AT recorded during the test possibly limited the expression of the productive potential of the animals. The lower production during winter is associated with higher energy demand for maintenance and lower feed digestibility (Arias *et al.*, 2008) due to increased activity of the thyroid gland that influences the gastrointestinal tract, causing increased intestinal motility and feed passage rate (NRC, 1996). In another study they mentioned that feed intake increased in winter with no effect on feed efficiency, so feed energy was directed to mitigate the effects of adverse weather conditions, such as increasing heat production or accumulation of body fat to help decrease heat loss (Mujibi *et al.*, 2010).

Serum Mineral Concentration

Zinc. Table 4 presents the concentration in blood serum, showing significance ($P<0.05$) on the day of sampling. Zn exceeded the highest level within the species, except on day 56, being this day where the three treatments registered the lowest concentration ($P<0.05$), with values of 11.14 ± 2.77, 10.56 ± 2.81 and 12.30 ± 2.90 pmol/L for T1, T2 and T3, respectively. Likewise, at the end of the experiment, T2 revealed a difference ($P<0.05$) showing 22.29 ± 2.81 pmol/L, lower at 1 and 28 d of sampling, but higher at 56 d. Reference values for Zn range from 13.96 to 16.43 pmol/L (McDowell and Arthington, 2005).

Zn concentration showed significance as levels increased at lower temperatures. The specific role of Zn in the immune response is not completely clear, however, it is considered essential for the integrity of the immune system (Hambridge *et al.*, 1986). Murray *et al.* (2000) mentioned that Zn is a structural component of the enzyme superoxide dismutase (SOD), which helps to scavenge free radicals produced in the body during an immune response. Droke and Spears (1993) published that the immune response of lambs fed marginally deficient zinc (8.7 mg/kg DM) in their diet revealed no differences from lambs that received

Table 4. Means (± SE) of zinc concentration (pmol/L) in blood serum of calves fed three different diets.

| Sampling day | T1 | Treatments[1] | | T3 |
		T2		
1	31.38 ± 2.77^{x}	31.65 ± 2.81^{x}		31.15 ± 2.90^{x}
28	30.03 ± 2.77^{x}	36.75 ± 2.81^{x}		29.62 ± 2.90^{x}
56	11.14 ± 2.77^{y}	10.56 ± 2.81^{z}		12.30 ± 2.90^{y}
84	29.05 ± 2.77^{x}	22.29 ± 2.81^{y}		24.76 ± 2.90^{x}

[x,y,z] Means with different column literals are different (P<0.05) between sampling days.

[1] T1 = Oat hay (HA) + corn silage (EM) + concentrate; T2 = HA + EM + concentrate + fermented apple bagasse (BMZN); T3 = HA + EM + concentrate + yeast inoculum (IL).

adequate amounts of zinc (44.0 mg/kg DM). However, in this study the lowest physiological Kmite was obtained on day 56 of sampling with no signs of disease, which is in agreement with Cuesta *et al.* (2011) who observed average values of 13.19 pmol/L lower than the lowest Kmite for the species with 87.5 % of the animals below the lower range with no physiological disease observed.

Spears (2000) reported that zinc deficiency in lambs decreases the percentage of lymphocytes and increases the number of neutrophils in the blood circulation; according to this author, in our study the decrease of Zn on day 56 led to a reduction in the number of lymphocytes, but not to an increase in the number of neutrophils, which coincides with that published by (Hambridge *et al.*, 1986). Cerone *et al.* (2000) reported that zinc and copper deficiency in cattle causes atrophy of the spleen and thymus, lymphopenia, monocytosis and decreased phagocytic capacity of neutrophils, affects T and B cells and therefore the production of antibodies.

Manganese. The highest (P<0.05) Mn concentration among the treatments was at the beginning and at the end of the test (Table 5), however, on day 28 the lowest concentrations (P<0.05) were observed with 4.97±2.15, 4.29±2.11 and 7.28±2.22 pmol/L for T1, T2 and T3, respectively. Except for T1 with similar levels on d^a 28 and 56 (P>0.05).

Table 5. Means (± SE) of manganese concentration (pmol/L) in blood serum of calves fed three different diets.

| Sampling day | T1 | Treatments[1] | | T3 |
		T2		
1	22.57 ± 2.15^{x}	21.10 ± 2.11^{x}		21.81 ± 2.22^{x}
28	4.97 ± 2.15^{y}	4.29 ± 2.11^{z}		7.28 ± 2.22^{z}
56	8.50 ± 2.15^{y}	12.56 ± 2.11^{y}		14.92 ± 2.22^{y}
84	22.30 ± 2.15^{x}	23.71 ± 2.11^{x}		23.37 ± 2.22^{x}

[x,y,z] Means with different column literals are different (P<0.05) between sampling days.

[1] T1 = Oat hay (HA) + corn silage (EM) + concentrate; T2 = HA + EM + concentrate + fermented apple bagasse (BMZN); T3 = HA + EM + concentrate + yeast inoculum (IL).

Mn absorption is influenced by many factors such as the chemical form and interactions between different micronutrients (Sanchez-Morito *et al.*, 1999). Likewise, Mn is another potential antagonist of Cu that affects its absorption at the intestinal level (Grace, 1973), limited studies have examined the antagonistic effect of Mn on Cu in ruminant diets. Arredondo *et al.* (2003) suggested that Mn and Cu may share the same pathway for intestinal absorption and transport, via divalent metal transporter protein 1, which makes them compete with each other. In their study Ivan and Grieve (1976) found that the addition of 50 mg Mn/kg DM to a diet containing 12 mg Mn/kg DM resulted in a decrease in Cu absorption in the gastrointestinal tract of Holstein calves; however, the dynamics of the antagonism in ruminants is not very clear.

In the present study, the decrease in Mn and Cu on sampling day 28 may have been due to the high concentration of Zn that occurred on the same day, with Cu on day 56 leading to the lowest amount of Mn and Zn.

Although there are other minerals that can cause its decrease such as Mg deficiency can indirectly modify Mn uptake by altering the availability of other divalent cations such as Ca, Zn (Planells *et al.*, 1993) and Cu (Jimenez *et al.*, 1997). Sanchez-Morito *et al.* (1999) assumed that the decrease in Mn concentration was related to an increase in metabolic activity in the bone marrow, as a mechanism to increase erythropoiesis and try to compensate for the haemolysis induced by Mg deficiency (Piomelli *et al., 1973*). On the other hand, Jenkins and Hiridoglou (1991) published that excess Mn negatively affects Fe metabolism in cattle, resulting in a decrease in cell pack volume and haemoglobin, reducing serum Fe binding capacity.

Copper. The Cu concentration of the three treatments exceeded the maximum level within the species during the test (Table 6), except on day 28, this same day all three treatments were lower ($P<0.05$) than the rest of the sampling days (8.04 ± 2.34, 8.32 ± 2.34 and 9.76 ± 2.47 pmol/L for T1, T2 and T3, respectively). However, the three treatments on day 56 revealed the highest ($P<0.05$) concentration than the rest of the sampling days, except for day 56 and 84 of T3 which were similar ($P>0.05$). The range of reference values for Cu in bovine serum is 11.0 to 18.0 pmol/L (McDowell and Arthington, 2005).

Castillo *et al.* (2012) mentioned that Cu is an essential trace element for immune response processes, it plays a role as a cofactor for many enzymes (cytochrome oxidase, ceruloplasmin and superoxide dismutase), so it can perform several functions in the immune system of which the direct mechanism of action is not very clear (Solaiman *et al.*, 2007). It has been observed that animals with Cu deficiency may have cupremia within the range, since in the tissues where it accumulates it continues to send its reserves to the circulation (Quiroz-Rocha and Bouda, 2001). Likewise, animals with chronic Cu intoxication may have excessive accumulation mainly in the liver and serum levels may be within reference ranges (Wikse *et al.*, 1992).

In this study, on day 28 of sampling, all three treatments showed levels below the physiological limits of the species, perhaps due to increased Zn competing with Cu uptake and affecting its concentration, or possibly because Cu was incorporated into SOD in red cells during the day of sampling.

Table 6. Means ($\pm$ SE) of copper concentration (pmol/L) in blood serum of calves fed three different diets.

Sampling day	T1	Treatments[1]	T3
		T2	
1	22.17 ± 2.34^y	24.48 ± 2.34^y	23.83 ± 2.47^y
28	8.04 ± 2.34^z	8.32 ± 2.34^z	9.76 ± 2.47^z
56	33.36 ± 2.34^x	38.06 ± 2.34^x	31.29 ± 2.47^x
84	23.16 ± 2.34^y	20.96 ± 2.34^y	29.29 ± 2.47^x

[x,y,z] Means with different column literals are different ($P<0.05$) between sampling days.

[1] T1 = Oat hay (HA) + corn silage (EM) + concentrate; T2 = HA + EM + concentrate + fermented apple bagasse (BMZN); T3 = HA + EM + concentrate + yeast inoculum (IL).

haematopoiesis and erythrocyte lifespan which is around 150 days (Suttle and McMurray, 1983). In another study, Cuesta *et al.* (2011) diagnosed average Cu values of 11.5 pmol/L and 40% of their cattle had values below the critical Kmite of 11.0 pmol/L without showing signs of clocosis. Most ruminants, except sheep, have a high capacity to store Cu in the liver (Mertz and Davis, 1987), this is due to their ability to synthesise metallothioneins which help to retain Cu and other minerals in the hepatocyte (Gooneratne *et al.*, 1989). During hypocuprosis, the activity of Cu-dependent enzymes such as cytochrome oxidase, which is necessary for phagocytic activity and superoxide dismutase (SOD), decreases, which reduces the half-life of leukocytes as these cells are dependent on this enzyme, causing a predisposition to infectious and viral d seases and negatively affecting the animal's immune system and thus its ability to respond to infection (Spears, 2003).

Antioxidant Activity

The highest AA (P<0.05) was observed at the beginning of the experiment in the three treatments, showing a significant increase (P<0.05) in T2 (15.99±0.03 pmol/L) when compared to T1 and T3, however, T1 on d^a 84 was lower (P<0.05) than T2 and T3 (15.62±0.03 pmol/L). Also, on d^a 28 and 56, T2 and T3 were similar to each other but lower (P<0.05) than d^a 1 and 84 of the test (Table 7). Tanaka *et al.* (2008) mentioned that heat stress stimulates the production of free radicals and reactive oxygen species (ROS). On the other hand, oxidative stress corresponds to an imbalance between the rate of oxidant production and degradation (Sorg, 2004), (Table 7). Tanaka *et al.* (2008) mentioned that heat stress stimulates the production of free radicals and reactive oxygen species (ROS). On the other hand, oxidative stress corresponds to an imbalance between the rate of oxidant production and degradation (Sorg, 2004).

The results of the present study showed that, when calves were exposed to fno in the first days of the experiment, the levels of minerals and BMZN in T2 exerted an effect on antioxidant enzymes, decreasing the release of ROS in calves, which may protect them from oxidative stress, showing the same pattern on day 84 for T2 and T3. This is in agreement with Prior and Cao (1999) who published that an increase in ROS production can cause a decrease in total antioxidant capacity *in vivo*. Rodriguez (2008) reported that sheep fattened on a manzarin diet showed an increase in AA with respect to the control group (24.34 and 21.79 pmol/L). In another study Gallegos (2007) published that Holstein cows in production fed with manzarin, showed higher AA (22.52 pmol/L) with respect to the control group (18.65 pmol/L) at the end of the test. Hahn and Mander (1997) reported that cattle require about 3 to 4 days after initiating a caloric shift to fully diminish the effects of the caloric load, followed by a recovery of the total antioxidant power in their body (Worapol *et al.*, 2011).

Hematic Biometry

Table 8 and 9 show haematological levels and reference parameters. Leukocytes (Leu), lymphocytes (Lin), mean corpuscular volume (MCV), mean corpuscular haemoglobin (MCH) and platelets (Pq) showed a significant increase in blood counts.

Table 7. Means (± SE) of antioxidant activity (pmol/L $_{FeSO4})_2$ in blood plasma of calves fed three different diets.

Sampling day	T1	Treatments[1] T2	T3
1	15.73±0.03[b x]	15.99±0.03[a x]	15.81±0.03[b x]
28	15.62±0.03[a y]	15.64±0.03[a z]	15.62±0.03[a z]
56	15.58±0.03[a y]	15.61±0.03[a z]	15.62±0.03[a z]
84	15.62±0.03[b y]	15.72±0.03[a y]	15.71±0.03[a y]

[ab] Means with different row literals are different (P<0.05) between treatments. [x,y,z] Means with different column literals are different (P<0.05) between sampling d^a.

[1] T1 = Oat hay (HA) + corn silage (EM) + concentrate; T2 = HA + EM + concentrate + fermented apple bagasse (BMZN); T3 = HA + EM + concentrate + yeast inoculum (IL).

[2] pmol/L FeSO4 (micromolar $_{FeSO4}$ reductive activity), per litre of blood plasma.

Table 8. Means (± SE)[2] of white blood cell biometry of calves fed three different diets.

calves fed three different diets

Cells	Sampling day	Treatment			Values of BH3
		T1	T2	T3	

Cells	Sampling day	T1	T2	T3	Value of BH[3]
Leukocytes (10^3 /pL)	56	7.58±0.95	8.18±0.95[y]	9.65±1.00[y]	4-12
	84	9.82±0.95 [x]	11.11±0.95[x]	10.80±1.00[y]	
Neutrophils (%)	56	3.22±0.76	6.67±0.76[a]	5.87±0.80[a]	15-45
	84	4.44±0.76 [a]	4.22±0.76[a]	5.00±0.80[a]	
Lymphocytes (%)	56	65.00±4.6	73.00±4.69	74.00±4.97	45-75
	84	74.44±4.69	76.89±4.69	71.50±4.97	
Mixed cells (%)	56	31.78±4.5	20.33±4.51	20.12±4.78	2-20
	84	21.11±4.5 [и]	18.89±4.51	23.50±4.78	
Neutrophils (10^3 /pL)	56	0.24±0.08	0.53±0.08[a]	0.61±0.09[a]	0.6-4
	84	0.43±0.08 [a]	0.48±0.08[a]	0.56±0.09[a]	
Lymphocytes (10^3 /pL)	56	4.84±0.92	5.99±0.92[y]	7.12±0.98[y]	2.5-7.5
	84	7.30±0.92 [x]	8.73±0.92[x]	7.75±0.98[y]	
Mixed cells (10^3 /pL)	56	2.49±0.40	1.65±0.40	1.91±0.42	0-2.4
	84	2.09±0.40	1.90±0.40	2.49±0.42	

[ab] Means with different row literals are different (P<0.05) between treatments. [xy] Means with different column literals are different (P<0.05) between sampling d^a.

[1]   T1 = Oat hay (HA) + corn silage (EM) + concentrate; T2 = HA + EM + concentrate + fermented apple bagasse (BMZN); T3 = HA + EM + concentrate + yeast inoculum (IL).

[2]   Means without row and column literals showed no difference (P>0.05).

[3]   Values of bovine haematic biometry.

Table 9. Means (± SE)2 of haematic red blood cell biometry of calves fed three different diets.

Cells	Sampling day	Treatment			Value of BH[3]
		T1	T2	T3	
Erythrocytes (106/pL)	56	5.14±0.41	5.40±0.41	5.32±0.44	5-10
	84	5.51±0.41	6.54±0.41	5.92±0.44	
Haemoglobin (g/dL)	56	11.45±0.50	11.33±0.50	11.72±0.53	
	84	11.20±0.50	12.62±0.50	11.60±0.53	8-15
Haematocrit (%)	56	38.42±2.41	34.02±2.41	33.77±2.56	
	84	33.60±2.41	37.87±2.41	34.80±2.56	24-46
MCV (fL)	56	73.78±3.26[x]	63.51±3.25[x]	64.09±3.45[x]	
	84	61.87±3.26[y]	59.69±3.25[x]	59.31±3.45[x]	40-60
HCM (Pg)	56	22.50±0.86[x]	21.15±0.86[x]	22.36±0.91[x]	
	84	20.17±0.86[y]	19.89±0.86[x]	19.77±0.91[y]	11-17
CHCM (g/dL)	56	31.47±1.07	33.29±1.07	35.35±1.14	
	84	33.30±1.07	33.30±1.07	33.30±1.14	30-36
Platelets (105/pL)	56	5.69±0.37[x]	5.67±0.37[x]	4.35±0.40[x]	
	84	2.41±0.37[y]	2.70±0.37[y]	2.06±0.40[y]	1-8

[xy] Means with different column literals are different (P<0.05) between sampling days.

[1]   T1 = Oat hay (HA) + corn silage (EM) + concentrate; T2 = HA + EM + concentrate + fermented apple bagasse (BMZN); T3 = HA + EM + concentrate + yeast inoculum (IL).

[2]   Means without row and column literals showed no difference (P>0.05).

3Hematic biometry values of cattle.

effect (P<0.05) of sampling day; neutrophils (Neu) in % and pL showed difference (P<0.05) between treatments.

Leukocytes and lymphocytes. The concentrations of Leu *and* Lin on day 84 of sampling showed significant increase (P<0.05) at T1 and T2 showing 9.82±0.95 and 11.11±0.95 x 10^3 /pL for Leu and 7.30±0.92 and 8.73±0 92 x 10^3 /pL for Lin, respectively. It is worth mentioning that T2 and T3 exceeded the within-species maximum level in Lin (Table 8). Calves that received IL in their diet maintained the amount of Leu and Lin close to the maximum level for these variables, showing no significant variation over time.

On day 56 of sampling, the lowest amount of Leu and Lin was observed, perhaps due to the presence of stress, provoked by the low average temperatures registered during days 1, 28 and 56 (-7.3, -3.0 and 0.9 °C) of the experiment followed by the decrease in the concentration of Zn, Mn and AA, which had repercussions in the decrease of these cells, exposing the calves to a release of ROS. However, on day 84 they increased the amount of white cells, minerals and AA, which may be related to the recovery of their comfort, overcoming the heat stress challenge without showing any adverse clinical signs.

Gupta *et al.* (2007) mentioned that there is a close relationship between leukocyte profile and plasma glucocorticoid level during physiological stress; these hormones may act by increasing the number and percentage of neutrophils, while decreasing lymphocytes (Blanco *et al.*, 2009). Buckham *et al.* (2008) reported that circulating lymphocytes adhere to endothelial cells lining the walls of blood vessels in response to increased glucocorticoids during stress, and subsequently move from the circulation to tissues such as lymph nodes, bone marrow, spleen and skin where they are retained, thereby producing a decrease in the number of circulating lymphocytes.

In another investigation, Rodriguez (2008) reported that sheep fattened with manzarine in their diet showed an increase in leucocytes (9.49 and 8.78 103/pL) at the end of the test. Gallegos (2007) reported that Holstein cows in production fed with and without manzarin had no effect (P>0.05) on the amount of leucocytes in blood.

Neutrophils. All three treatments during the test were below the lower limit of Neu in percentage and concentration, except T3 at d^a 56. Despite this behaviour, T1 presented the lowest (P<0.05) amount of Neu on d^a 56 (3.22±0.76 % and 0.24±0.08 x 10^3 /pL), with respect to the rest of the treatments (Table 8).

The low amount of Neu may have been due to the low concentration of Zn and Mn on day 56, without reaching the optimum amount at the end of the test despite the increase of the minerals mentioned. On the other hand, even though they were below the lower Kmite, T1 on day 56 was the most detrimental with respect to T2 and T3. In addition, it was observed that T2 and T3 animals showed a higher level of Neu than T1 during the test. Rodriguez (2008) reported a sex effect in sheep fed manzarin, where females that received manzarin had higher Neutrophil counts than males (59.2 and 50.2 %, respectively) in the control group. On the other hand, Gallegos (2007) found no difference in Holstein cows in production fed with and without manzarin (34.27 and 35.98 %).

Romero *et al.* (2011) mentioned that neutrophils proliferate in the circulation as a response to infections, inflammation and stress; decreasing in certain infections and states of anaphylaxis. On the other hand, glucocorticoids in stressed animals stimulate the flow of neutrophils from the bone marrow into the blood and reduce their passage to other compartments, leading to an increase of mature and immature neutrophils in the blood circulation (Buckham *et al.*, 2008).

Mean corpuscular volume. The MCV values of the three treatments on day 56 exceeded the maximum level within the species, likewise, the MCV of T1 on day 84, although it was lower (61.87±3.26 fentolitres (fL)) on day 56, exceeded the maximum level (Table 9).

The high MCV observed on day 56 and 84 of the test at T1 may have been detrimental to the low amounts of Zn, Mn and AA, which may have increased an inflammatory reaction and macrocytic anaemia. Increased MCV is an indicator of macrocytic anaemia, while eosinophils decrease in response to increased inflammatory reaction (Rodostitis *et al.*, 2000). In this study there may have been a decrease in Fe concentration due to the effect of the high levels of Mn and Cu shown in this investigation, which is consistent with the findings of Reeves *et al.* (2004). Likewise, Jenkins and Hiridoglou (1991) mentioned that high Mn concentrations affect Fe metabolism in cattle, resulting in a decrease in cell pack volume and haemoglobin, which reduces serum Fe binding capacity. On the other hand, Rodriguez (2008) found a similar behaviour when manzarin was added to the sheep diet, where the supplemented animals decreased the amount of VCM compared to the control group (32.99 and 33.57 fL) but without deviating from the normal values within the species. In another study, Coppo and Mussart (2006) found no significant effect of supplementing heifers with citrus pomace in winter for 90 days compared to the control group (46.5 and 47.0 fL).

Mean corpuscular haemoglobin. The MCH levels of the three treatments on the two sampling days exceeded the maximum reference level (Table 9), on day 84 T1 and T3 decreased (P<0.05) with respect to day 56 (20.17±0.86 and 19.77±0.91 picograms (Pg), respectively), but exceeded that reported by Aiello and Mays, 2000. There are reports that MCH indicates haemoglobin concentration based on red blood cell count, similar to that indicated by mean corpuscular haemoglobin concentration (MCHC), although the latter is considered more appropriate. Rodriguez (2008) found significant differences between sheep sex and sampling, where males were superior to females (12.74 and 12.10 Pg), as well as exceeding the highest range within the species. In another study, Coppo and Mussart (2006) found a significant effect of supplementing heifers with citrus pomace in winter for 90 d with respect to the control group (16.6 and 14.5 fL, respectively).

Platelets. The amount of Pq of the three treatments (Table 9) was higher (P<0.05) on d^a 56 with respect to d^a 84 (5.69±0.37, 5.67±0.37 and 4.35±0.40 x 10%iL for T1, T2 and T3, respectively). The Pqs showed sampling effect, the highest amounts were observed on day 56, which could be due to the increase of antibodies observed during the test, as indicated by other studies. Aiello and Mays (2000) mentioned that the decrease in platelet production may be due to drugs, toxins and in some cases immunological activity, due to the production of antibodies that can bind to the platelet surface. All three treatments showed the lowest Pq on day 84, perhaps due to the increased number of white blood cells observed. Rodriguez (2008) reported an effect between sex of sheep and sampling, where males in the second sampling were higher than females showing values of 7.6 and 5.4 x 103/pL. On the other hand, Gallegos (2007) found no difference (P>0.05) in the amount of platelets when supplementing Holstein cows in production with and without manzarin (4.5 and 3.8 x 10^3 /pL).

CONCLUSIONS AND RECOMMENDATIONS

Under the conditions in which this work was carried out, it is concluded that the addition of a yeast inoculum in the diet of growing calves increased the CDA and CA.

At low ambient temperatures, the highest concentration of Zn in serum was observed, with Mn and Cu decreasing on day 28. Likewise, calves receiving BMZN and IL had higher AA at the end of the test. We suggest that this behaviour was due to the fact that they favoured antioxidant enzymes and therefore decreased ROS release.

Leukocytes, lymphocytes, MCV, MCH and platelets showed differences across sampling days. Also, T1 showed lower Neu in % and pL on day 56. These changes indicate that the thermal stress occurred in the animals, which decreased the number of white cells and increased the number of red cells.

Further investigation with BMZN and IL supplementation is recommended, exposing early weaned calves to less stressful temperatures.

LITERATURE CITED

Aiello, S. E. and A. Mays. 2000. The Merck Veterinary Manual. 5th ed. Oceano Grupo Editorial, S. A. Spain.

Aksu, T., B. Ozsoy, D. S. Aksu, M. A. Yoruk and M. Gul. 2011. The effects of lower levels of organically complexed zinc, copper and manganese in broiler diets on performance, mineral concentration of tibia and mineral excretion. Kafkas Univ Vet Fak Derg. 17: 141-146.

Arias, R. A. 2006. Environmental factors affecting daily water intake on cattle finished in feedlots. Master Thesis, University of Nebraska-Lincoln, Nebraska, USA.

Arias, R. A., T. L. Mader and P. C. Escobar. 2008. Climatic factors affecting the productive performance of beef and dairy cattle. Arch. Med. Vet. 40:7-22.

Arredondo, M., P. Munoz, C. V. Mura and M. Nunez. 2003. DMT1, a physiologically relevant apical Cu^{1+} transporter of intestinal cells. Am. J. Physiol. Cell Physiol. 284:C1525-C1530.

Becerra, A., C. Rodriguez, J. Jimenez, O. Ruiz, A. EKas and A. Rammez. 2008. Urea and corn in the aerobic fermentation of apple bagasse for protein production. TECNOCIENCIA Chihuahua. 2:7-14.

Benzie, I. F. F. and J. J. Strain. 1996. Ferric reductive ability of plasma (FRAP) as a measure of antioxidant power: the frap assay. Anal. Biochem. 239:7076.

Bm^k, H. and I. I. Turkmen. 2001. The effect of *Saccharomyces cerevisiae* on *in vitro* rumen digestibilities of dry matter, organic matter and neutral detergent fiber of different forage: concentrate ratios in diets. J. Fac. Vet. Med. 20:29-37.

Blanco, M., I. Casasus and J. Palacic. 2009. Effect of age at weaning on the physiological stress response and temperament of two beef cattle breeds. Animal. 3:108-117.

Buckham Sporer, K. R., P. S. D. Weber, J. L. Burton, B. Earley and A. Crowe. 2008. Transportation of young beef bulls alters circulating physiological parameters that may be effective biomarkers of stress. J. Anim. Sci. 86:1325-1334.

Castillo, C., J. Hernandez, M. Garda Vaquero, M. Lopez Alonso, V. Pereira, M. Miranda, I. Blanco and J. L. Benedito. 2012. Effect of moderate Cu supplementation on serum metabolites, enzymes and redox state in feedlot calves. Vet Res. Commun. 93.269-274.

Cerone, S. I., A. S. Sansinanea, S. A. Streitenberg, M. C. Garrta and N. J. Auza. 2000. Bovine monocyte-derived macrophage function in inducing copper deficiency. Gen Phy Biophys. 19:49-58.

Chew, P. B. 1995. Antioxidant vitamins affect food animal immunity and health. J. Nutr. 125: 1804S-1808S.

Coppo, J. A. and N. B. Mussart. 2006. Citrus pomace as a winter supplement for zebu crossbred heifers in Argentina. Vol. VII, No. 04, April.

Cuesta, M. M., D. J. R. Garaa, P. E. A. Silveira and G. Y. Pino. 2011. Parenteral administration of a zinc, copper and manganese compound in dairy cows. Electronic Journal of Veterinary Medicine. 12:28-36.

Czarnecki-Maulden, G. L. 2008. Effect of dietary modulation of the intestinal microbiota on reproduction and early growth. Theriogenology 70:286-290.

Delfino, J. G. and G. W. Mathison. 1991. Effects of cold environment and intake level on the energetic efficiency of feedlot steers. J. Anim. Sci. 69:45774587.

Desnoyers, M., S. Giger-Reverdin, G. Bertin, C. Duvaux-Ponter and D. Sauvant. 2009. Metaanalysis of the influence of *Saccharomyces cerevisiae* supplementation on ruminal parameters and milk production of ruminants. J. Dairy Sci. 92:1620-1632.

D^az, P. D. 2006. Production of microbial protein from waste apples added with urea and soybean paste. Master's thesis. Faculty of Animal Husbandry. Autonomous University of Chihuahua. Chihuahua. Chih. Mex.

Dorton, K. L., T. E. Engle, R. M. Enns and J. J. Wagner. 2007. Effects of trace mineral supplementation, source and growth implants on immune response of growing and finishing feedlot steers. The Professional Animal Scientist. 23:29-35.

Droker, A. E., and J. W. Spears. 1993. *In vitro* and *in vivo* immunological measurements in growing lambs fed diets deficient, marginal or adequate in zinc. J. Nutr. Immunol. 2:71-90.

Gallegos, A. M. A. 2007. Somatic cell count in milk, plasma antioxidant activity and blood cellular components of producing holstein cows fed manzarin in the diet. Doctoral dissertation. Faculty of Animal Science and Ecology. University
Autonoma de Chihuahua. Chihuahua, Chih. Mex.

Gooneratne, S. R., W. T. Buckley and D. A. Christensen. 1989. Review of copper deficiency and metabolism in ruminants. Can J Anim Sci. 69:819-845.

Grace, N. D. 1973. Effect of high dietary Mn levels on the growth rate and the level of mineral elements in the plasma and soft tissues of sheep. New Zeal J. Agric. Res. 16:177-180.

Gupta, S., B. Earley and M. A. Crowe. 2007. Effect of 12-hour road transportation on physiological, immunological and hematological parameters in bull housed at different space allowances. Vet. J. 173:605-616.

Hambridge, K. M., C. E. Casey and N. F. Krebs. 1986. Zinc. In trace elements in human and animal nutrition. 5th ed. Walter Mertz. Academic Press. Inc., London. United Kingdom.

Huerta, J. M., M. E. C. Ortega and M. P. Cobos. 2005. Oxidative stress and the use of antioxidants in domestic animals. Journal of Science and Technology of America. 30:728-734.

INAFED. 2010. National Institute for Federalism and Municipal Development. http://www.inafed.gob.mx/work/templates/enciclo/chihuahua/. Accessed September 10, 2012.

Ivan, M. and C. M. Grieve. 1976. Effects of zinc, copper and manganese supplementation of high-concentrate ratio non gastrointestinal absorption of copper and manganese in Holstein cows. J. Dairy Sci. 59:1764-1768.

Jenkins, K. J. and M. Hidiroglou. 1991. Tolerance of the preruminant calf for excess manganese or zinc in milk replacer. J. Dairy Sci. 74:1047-1053.

Jimenez, A., E. Planells, P. Aranda, M. Sanchez-Vinas and J. Llopis. 1997. Changes in bioavailability and tissue distribution of copper caused by magnesium deficiency in rats. J. Agric. Food Chem. 45:4023-4027.

Makino, T. and K. Takahara. 1981. Direct determination of plasma copper and zinc in infants by atomic absorption with discrete nebulization. Clin. Chem. 27:1445-1447.

McDowell, L. R. and J. D. Arthington. 2005. Minerals for Grazing Ruminants in Tropical Regions. 4th ed. Dep. Zoot. University of Florida. Gainesville. IFAS. U. S. A.

Mertz, W. and G. K. Davis. 1987. Copper. In: Mertz W, editor. Trace Elements in Human and Animal Nutrition. Vol I. 5$^{th\ ed.}$ Philadelphia: Academic Press. U.S.A.

Moberg, G. P. 1987. A model for assessing the impact of behavioral stress on domestic animals. J. Anim. Sci. 65:1228-1235.

Mujibi, F. D. N., S. S. Moore, D. J. Nkrumah, Z. Wang and J. A. Basarab. 2010. Season of testing and its effect on feed intake and efficiency on growing beef cattle. J. Anim. Sci. 88: 3789-3799.

Murray, R. K., D. K. Granner, P. A. Mayes, and V. W. Rodwell. 2000. W. Rodwell. 2000. Harper's Biochemistry. 25th ed. p 135, 223. McGraw Hill Health Professional Division, New York. U.S.A.

Newbold, C. J., R. J. Wallace and F. M. McIntosh. 1996. Mode of action of the yeast *Saccharomyces cerevisiae* as a feed additive for ruminants. British J. Nutr. 76:249-251.

NRC. 1996. Nutrient Requirements of Beef Cattle. 7th ed. National Academy Press, Washington, DC. U.S.A.

Piomelli, S., V. Jansen and F. Dancis. 1973. The haemolytic anaemia of magnesium deficiency in adult rats. Blood. 41:451-459.

Planells, E., P. Aranda, A. Lerma and J. Llopis. 1994. Changes in bioavailability and tissue distribution of zinc caused by magnesium deficiency in rats. Brit. J. Nutr. 72:315-323.

Prior, R. L. and G. Cao. 1999. *In vivo* total antioxidant capacity: comparison of different analytical methods. Free Radic Biol Med. 27:1173-1181.

Quiroz-Rocha, G. F. and J. Bouda. 2001. Physiopathology of copper deficiencies in ruminants and their diagnosis. Vet. Mex. 32:289-296.

Reeves, P. G., N. V. C. Ralston, J. P. Idso and H. Lukaski. 2004. Contrasting and cooperative effects of copper and iron deficiencies in male rats fed different concentrations of manganese and different sources of sulfur amino acids in an AIN-93G-based diet. J. Nutr. 134:416-425.

Rodostitis, O. M., C. C. Gay, D. C. Blood, K. Gay, D. C. Blood and K. W. Hinchliff. 2000. Veterinary Medicine, 9th ed. Harcourt Publisher, London. United Kingdom.

Rodriguez, R. H. E. 2008. Obtaining Manzarin from Apple By-Products and its Effect on the Health of Fattening Lambs. Doctoral Dissertation. Faculty of Animal Science and Ecology. Autonomous University of Chihuahua. Chihuahua. Chih. Mex.

Rodriguez-Muela, C., D. D^az, F. Salvador, O. Ruiz, C. Arzola, A. Flores, O. La O and A. EKas. 2010. Effect of urea and soybean paste levels on protein concentration during solid-state fermentation of apple (*Malus domestica*). Rev. Cubana de Cien. Agric, 44:23-26.

Romero, P. M. H., L. F. Uribe-Velasquez and V. J. A. Sanchez. 2011. Biomarkers of stress as indicators of animal welfare in beef cattle. Biosalud. 10:71-87.

Sanchez-Morito, N., E. Planells, P. Aranda and J. Llopis. 1999. MagnesiumManganese Interactions Caused by Magnesium Deficiency in Rats. J. Am. Coll. Nutr. 18:475-480.

SAS. 2004. SAS/STAT® 9.1 User's Guide. SAS Institute Inc. United States of America.

Solaiman, S. G., T. J. Craig Jr., G. Reddy and C. E. Shoemaker. 2007. E. Shoemaker. 2007. Effect of high levels of Cu supplement on growth performance, rumen fermentation, and immune responses in goats kids. Small Rumin. Res. 69:115-123.

Sorg, O. 2004. Oxidative stress: a theoretical model or a biological reality? C. R. Biol. 327:649-662.

Spears, J. W. 2000. Micronutrinet and immune function in cattle. Proceedings of the Nutrition Society. 59:587-594.

Spears, J. W. 2003. Trace mineral bioavailability in ruminants. J. Nutr. 133:15061509.

Steel, R. G. D. and J. H. Torrie. 1997. Biostatistics. Principles and Procedures. 2nd ed. McGraw-Hill. Mexico.

Suttle, N. F. and C. H. McMurray. 1983. Use of erythrocyte copper-zinc superoxide dismutase activity and hair or free concentrations in the diagnosis of hypocuprosis in ruminants. Res. Vet. Sci. 35:47-52.

Tanaka, M., Y. Kamiya, T. Suzuki, M. Kamiya and Y. Nakai. 2008. Nakai. 2008. Relationship between milk production and plasma concentrations of oxidative stress markers during hot season in primiparous cows. J. Anim. Sci. 79:481-486.

van der Peet-Schwering, C. M. C., A. J. M. Jansman, H. Smidt and I. Yoon. 2007. Effects of yeast culture on performance, gut integrity, and blood cell composition of weanling pigs. J. Anim. Sci. 85:3099-3109.

Villagran, D., Rodriguez-Muela, C., Burrola, E., Gonzalez, E., Ortega, A. 2009. Identification of yeasts involved in the solid-state fermentation of apple bagasse. Page 2 in Proceedings of the XLV Meeting. National Livestock Research Institute. Saltillo, Coah, Mex.

Wikse, S. E., D. Herd, R. Field and P. Holland. 1992. Diagnosis of copper deficiency in cattle. J. Anim. Vet. Med. Assoc. 200:1625-1629.

Worapol, A., K. Watee and B. Thongchai. Thongchai. 2011. Effects of shade on physiological changes, oxidative stress and total antioxidant power in Thai Brahman cattle. Int. J.

Biometeorol. 55:741-748.

Young, B. A. 1983. Ruminant cold stress: Effect on production. Can. J. Anim. Sci. 57:1601-1607.

Young, B. A., B. Walker, A. E. Dixon and V. A. Walker. 1989. Physiological Adaptation to the Environment. J. Anim. Sci. 67:2426-2432.

IN VITRO FERMENTATION OF DIETS WITH A
YEAST INOCULUM
AND FERMENTED APPLE BAGASSE FOR
GROWING CALVES

SUMMARY

IN VITRO FERMENTATICN OF DIETS WITH THE ADDITION OF A
YEAST INOCULUM
AND FERMENTED APPLE BAGASSE FOR CALVES IN
GROWTH

The objective was to evaluate the effect on *in vitro* fermentation of diets with the addition of a yeast inoculum and fermented apple bagasse for growing calves. The treatments consisted of: T1 (control): oat hay (HA) + corn silage (EM) + concentrate; T2: HA + EM + concentrate + apple bagasse (BMZN) and T3: HA + EM + concentrate + yeast inoculum (IL). The variables evaluated were dry matter (DM) digestibility, neutral detergent fibre (NDF), acid detergent fibre (ADF) and acid detergent lignin (ADL) content, gas production volume, concentration of acetic, propionic and butmic acid, ammoniacal nitrogen (N-NH3), lactic acid and pH. *In vitro* digestibility was carried out at 48 h, while for the rest of the variables at 3, 6, 12, 24, 48, 72 and 96 h. A completely randomised design was used, the fibre digestibility variables were analysed with a statistical model including treatment as a fixed effect, the concentration of volatile fatty acids, N-NH3, lactic acid and pH were analysed with a model including treatment and time as fixed effects. For gas production, treatment, time and their interaction were considered as fixed effects. To analyse parameters A, B and C of the fitted non-linear regression model, treatment was used as a fixed effect. The highest digestibility ($p<0.05$) of DM (72.02 and 72.49 %), NDF (70.77 and 70.25 %), FDA (64.07 and 64.56 %) and the lowest LDA content (4.25 and 4.14) was shown by T2 and T3. T3 was superior ($p<0.05$) in gas production volume with values of 4.20, 5.93, 6.73 and 7.33 mL/0.2 g DM at 24, 48, 72 and 96 hours, respectively. The highest concentration ($p<0.05$) of VFA was presented by T2 and T3 (with values for acetic, propionic and butmic of 16.00 and 17.19, 6.54 and 6.13, 2.63 and 2.67 mmol/L, respectively). T2 and T3 showed an increase ($p<0.05$) at the end of *in vitro* fermentation in N-NH3 concentration (0.21 and 0.22 mM/mL) and a decrease ($p<0.05$) in lactic acid (1.46 and 1.37 mM/mL), while T3 showed an increase ($p<0.05$) in pH (6.74). It is concluded that the addition of BMZN and IL to the diet of growing calves favours DM, NDF and FDA digestibility, decreases LDA content, increases VFA and N-NH3 concentration and causes a decrease in lactic acid.

INTRODUCTION

The nutritive value of feedstuffs is determined by the bioavailability of nutrients and the dynamics of solubility processes in the gastrointestinal tract. Van Soest (1994) mentioned that the cell wall is the major organic constituent of forages, comprising 40 to 80 % of the dry matter and consisting of structural polysaccharides such as cellulose, hemicellulose and lignin. However, fibre digestibility is limited by the degree of maturity and lignification of the forages (Akin, 1989). Various strategies have been used to improve the digestibility of forages, such as agronomic advances and forage crossbreeding programmes (Beauchemin *et al.*, 2003).

In previous years, yeast cultures have been used to improve the nutritive value and efficient utilisation of low-quality pasture. Additions of yeast cultures to ruminant diets can increase dry matter intake, productive performance, cellulose degradation and nutrient digestibility (Lesmeister *et al.*, 2004). *Saccharomyces cerevisiae* (Sc) has been widely used as a supplement in ruminant diets. Benefits associated with Sc include increased digestion of neutral detergent fibre (NDF) and dry matter (DM; Plata *et al.*, 1994), increased initial rate of fibre digestion (Williams *et al.*, 1991), improved microbial efficiency and *in situ* crude protein (CP) degradation (Olson *et al.*, 1994).

Bruni and Chilibroste (2001) mentioned that as an alternative to substrate disappearance, cumulative gas production is measured as an indicator of carbon metabolism, focusing on the accumulation of fermentation end-products such as carbon dioxide (CO_2), methane (CH_4) and volatile fatty acids (VFA). However, the yeasts *Kluyveromyces lactis* and *Issatchenkya orientalis* have been little studied as additives to improve the digestibility of fibrous diets in ruminants.

Therefore, the objective was to evaluate the effect of a yeast inoculum and fermented apple bagasse containing *Kluyveromyces lactis*, *Issatchenkya orientalis* and *Saccharomyces cerevisiae* added to the diet of weaned calves on fibre digestibility, gas production and VFA profiles. The results of the present study will allow producers and livestock nutritionists to have a better understanding of the use of these feed additives on fibre digestibility and VFA profile, allowing to optimise their use in this type of animals.

MATERIALS AND METHODS

Location of the Study Area

This research was carried out in the facilities of the Facultad de Zootecnia y Ecolog^a of the Universidad Autonoma de Chihuahua, Chih., Mexico, located in the coordinates 28° 35' 07" north latitude and 106° 06' 23" west longitude, with an altitude of 1,517 masl, according to the Global Positioning System (GPS, MAGELLAN® , MobileMapper-Pro). The average annual temperature is 18.2 °C, with an average maximum temperature of 37.7 °C and a mean annual temperature of 18.2 °C.

an average minimum of -7.4 °C. The average annual precipitation is 387.5 mm, with 71 rainy days per year and a relative humidity of 49 %; predominantly an extreme semi-arid climate (INAFED, 2008).

Description of Treatments

For the development of this experiment, the three diets offered to growing calves as mentioned in Experiment I were used. T1 (control): oat hay (HA) + corn silage (EM) + concentrate; T2: HA + EM + concentrate + fermented apple bagasse (BMZN); T3: HA + EM + concentrate + yeast inoculum (IL).

Chemical Analysis of Diets

In order to obtain the composite values of dry matter (DM) digestibility, neutral detergent fibre (NDF), acid detergent fibre (ADF) and acid detergent lignin (ADL) content, the chemical composition of the above mentioned diet fractions was analysed. Samples of the diets were collected every 15 d to form monthly composite samples. Samples were dried at 60 °C for 48 h in a forced-air oven and ground to 1 miKmeter (mm) in a Wiley mill (Arthur H. Thomas Co., Philadelphia, PA), these samples were dried at 105 °C for 8 h in a forced-air oven to determine absolute DM and sequentially incinerated at 600 °C for 4 h in a muffle to determine ash (AOAC, 2000). In the composite samples of forage and concentrate, the concentration of NDF, FDA (Van Soest *et al.,* 1991) and LDA (Goering and Van Soest, 1970) were determined. For NDF analysis, sodium sulphite (Na_2SO_3) and thermostable a-amylase (Ankom Technology) were used to remove nitrogenous matter and starch.

In vitro digestibility of DM, NDF, FDA and LDA content of the diet.

Prior to incubation, the bags were immersed in acetone to remove the surfactant that inhibits microbial digestion, and then dried at room temperature. Next, the composite samples of the diets (0.45 g ± 0.05) were weighed into ANKOM® F57 filter bags (25 pm pore size and dimensions of 5 x 4 cm) identified ard heat sealed and incubated *in vitro* according to the Daisy methodology[11] (ANKOM Technology Corp., Fairport, NY-USA), which involves buffer solutions A and B and ruminal inoculum.

The culture medium was prepared according to the procedure of ANKOM Technology® . Buffer A was composed of 10 g/L KH_2PO_4, 0.5 g/L $MgSO^7H_2O$, 0.5 g/L NaCl, 0.1 g/L CaC^2H_2O and 0.5 g urea in one litre of distilled water. Buffer solution B consisted of 15 g Na_2CO_3 and 1.0 g $Na_2S^9H_2O$ in one litre of distilled water. To prepare the inoculum, the two solutions were preheated to a temperature of 39 °C and mixed in a 5:1 ratio (1330 mL solution A and 266 mL solution B) adjusting the pH to 6.8 at 39 °C. To this mixture of solutions, 400 mL of ruminal Kquido was added to obtain an amount of 2000 mL, which was distributed in the four jars of the *DAISYII* digester *(500* mL per jar), subsequently, the bags of each sample were placed in triplicate, adding a standard and a blank bag (without sample) and proceeded to the analysis of true digestibility *in vitro* for 48 h at 39 °C (± 0.5).

After incubation, the bags were removed from the jars and washed with tap water until the water ran clear. The bags were dried in a forced air oven for three hours at 105 °C to determine DM digestibility. Subsequently, NDF and FDA were determined sequentially on the ANKOM apparatus[200] (Ankom Technology Corp., Fairport, NY) according to the methodology described by Van Soest *et al.* (1991) and LDA following the procedure of Goering and Van Soest (1970).

Rumen Fluid Collection

Rumen fluid was collected 15 min prior to feeding, using two Herford cows with cannulae as inoculum donors. These animals were fed a corn silage based maintenance diet with fresh water freely available. The rumen inoculum was acquired directly from the rumen, with the help of 3 folds of gauze to collect rumen content and deposit it in a thermic container (2 L) at 39 °C, adding a small amount of rumen digesta from each animal. It was then transported to the Animal Nutrition laboratory, where it was filtered through two layers of filter cloth, removing the solid part of the cloth and depositing it in a blender together with ruminal Kquido, liquefying them for 30 s, applying carbon dioxide (CO_2) constantly to guarantee anaerobic conditions.

The liquefied solution together with the remaining ruminal Kquido was filtered again and deposited into a vessel kept in a water bath at 39 °C and continuously saturated with CO_2. This procedure ensures that the inoculum is composed of microorganisms in the kieselguhr and in the fibre. Finally, a total of 400 mL of ruminal inoculum was deposited into each of the digestion jars, where the mixed buffer solutions were gassed for 30 s with CO_2.

In vitro gas production

In vitro gas production (PG) was performed using the protocol of Menke and Steingass (1988), with the modifications mentioned by Muro (2007). This technique is widely used to determine the amount of gas produced from a feed in an incubation period, which is related to the degradation of the feed.

The sample arrangement involves grinding the substrate on a 1 mm mesh (Menke *et al.*, 1979). The *in vitro* rumen medium consisted of solution A (micromineral) with 13.2 g CaChH2O, 10.0 g MnC^4H2O, 1.0 g CoC^6H2O and 8.0 g FeCls^6H2O gauged in 100 mL of distilled water; solution B (solution bufer) with 39.0 g NaHCO3 gauged in 1 L of distilled water; solution C (macromineral) with 5.7 g Na2HPO4, 6.2 g KH2PO4 and 0.6 g MgSO^7H2O gauged in 1 L of distilled water; rezasurin solution (100 mg rezasurin) gauged in 100 mL of distilled water, used as an indicator of anaerobiosis; and finally the reducing solution with 4.0 g 1N NaOH and 0.625 g Na2S^9H2O gauged in 100 mL of distilled water.

Solutions A, B, C and resazurin were added to a flask with distilled water, which was placed in a rotary Shaker incubator at 39 °C, then the reducing solution was added and gassed with CO2 until the colour of the mixture (artificial saliva) turned from blue to light pink. The incubations of the samples were carried out in triplicate plus one blank in each of the hours, using 50 mL glass bottles, sealed with rubber stoppers; to these were added 0.2 g of sample, 10 mL of ruminal Kquido and 20 mL of artificial saliva. The bottles with sample, ruminal fluid and artificial saliva were sealed and placed in the incubator (Shaker I2400) at 39 °C with constant agitation (67 revolutions per minute; rpm), protected from light during the 96 h of the test. The internal pressure of the flasks due to food degradation was measured with a pressure transducer (FESTO®) at 3, 6, 12, 24, 48, 72 and 96 h of incubation, by puncturing the flasks, and the accumulated pressure was recorded (Theodorou *et al.*, 1994). The results were expressed in mL PG per 0.2 g DM (mL PG/0.2 g DM).

AGV Production and Profiling

A 20 mL sample was collected from the PG flasks for determination of acetic, propionic and butmco acids at 3, 6, 12, 24, 48, 72 and 96 h, which were filtered with two layers of gauze to separate the solid and kieselguhr contents. The pH of the latter was determined immediately, using a potentiometer (Combo, HANNA instruments® Inc., Woonsocket, RI), and a 10 mL subsample was taken and centrifuged at 3,500 *xg at* 4 °C for 10 min. The supernatant was deposited in previously marked amber vials, acidified with 0.2 mL of 50 % H2SO4 and frozen at - 20 °C until analysis. Prior to measurement, the samples were thawed in refrigeration at 4 °C, added with 25 % metaphosphoric acid and centrifuged again with the aforementioned characteristics and preserved in refrigeration until analysis by gas chromatography (Brotz and Schaefer, 1987).

Volatile fatty acids (VFA) were analysed by means of a gas chromatograph (SRI 8610, SRI Instruments, CA), with an Alltech ECONO-CAPTM ECTM column of the following dimensions: 15 m long, external diameter (0.53 mm) and 0.25 mm internal diameter. 0.6 pL was injected at an injector temperature of 230 °C and flame detector 250 °C, the oven temperature ramp was 100 °C, 20 °C per min and 190 °C for 1 min, with an average reading time of 1.51, 1.92 and 2.48 min for acetic acid, propionic acid and butmic acid, and a net reading time of 6.83 min.

Ammoniacal Nitrogen (N-NH3)

A 20 mL sample was obtained from the PG flasks to analyse N-NH3 at 3, 6, 12, 24, 48, 72 and 96 h, which were filtered with two layers of gauze to separate the solid material and kieselguhr; the pH was determined in the same way as the VFA sample. A 10 mL subsample was then collected and centrifuged in the same way as the VFA sample. The supernatant was then decanted into previously labelled 20 mL plastic containers, frozen at - 5 °C for storage until analysis, thawed at 4 °C prior to measurement and the concentration determined by colorimetry according to the technique of Broderick and Kang (1980).

The prediction equation to calculate the N-NH3 concentration was obtained from the triplicate analysis of standard solutions with levels of 0, 5, 10, 15, 15, 20 and 25 pL of N-NH3/mL, using distilled water as blank at the zero point (0 pL). The procedure to measure the absorbance of the samples, the standard solution and the blanks, was developed as follows: 920 pL of distilled water and 80 pL of the concentrated sample were added in test tubes to complete 1 mL, they were mixed in vortex (Vortex Genie II) later 50 pL of the diluted sample were taken for each repetition and deposited in test tubes, adding 2.5 mL of phenol was added and mixed in vortex, then 2 mL of hypochlorite was added, mixing again; for the standard or blank 50 pL of distilled water, 2.5 mL of phenol and 2 mL of hypochlorite were added, the obtained solutions were mixed in vortex and incubated for 5 min in a water bath (90 to 100 °C).

The samples were then allowed to cool for 5 min at room temperature and the absorbance of each sample was measured at a wavelength of 630 nanometres (nm), previously the absorbance value was adjusted to 0 using blanks as a reference. With the results obtained from the absorbance, the prediction equation obtained with the standard solution and the dilution percentage of the samples, the concentration of N-NH3 in millimolar per millilitre of sample (mM/mL) was calculated.

Lactic acid

A 20 mL sample was collected from the PG bottles for lactic acid analysis at 3, 6, 12, 24, 48, 72 and 96 h; the sample handling procedure was similar to the N-NH3 variable up to the time of analysis. Lactic acid concentration was determined by colorimetry according to the procedure of Taylor (1996). For the standard, 5, 10, 15, 20 and 25 pL of lactic acid/mL and distilled water as blank (0 pL lactic acid/mL) were used.

The analysis was performed in triplicate, 3 mL of concentrated H2SO4 and 0.5 mL of the diluted sample were added in test tubes of the standard (blank), the solutions obtained were vortex mixed (Vortex Genie II) and incubated for 10 min in a water bath (95 to 100 °C).

Subsequently, 100 pL of 4 % CuSO4 solution and 200 pL of 1.5 % p-phenylphenol solution in 95 % ethanol were added to each tube, mixed again and allowed to stand for 30 min at room temperature (not less than 20 °C). The absorbance of the contents of each tube was measured at a wavelength of 570 nm. Using the lactic acid concentration and the prediction equation obtained with the standard solutions and the dilution percentage of the samples, the concentration of lactic acid in millimolar per millilitre of sample (mM/mL) was calculated.

Statistical Analysis

For the analysis of DM, NDF, FDA and LDA digestibility, a statistical model was used that included treatment as a fixed effect in a completely randomised design. On the other hand, for the variables of acetic acid, propionic acid, butmic acid, N-NH3, lactic acid and pH, a statistical model including treatment and time as fixed effect was used. Also, gas production was analysed with a similar model that included treatment, time and their interaction as fixed effects. To analyse parameters A, B and C of the fitted non-linear regression model, treatment was used as a fixed effect. Tukey's test was used to establish possible differences between treatment means (Steel and Torrie, 1997).

The *in vitro* accumulated PG amounts were fitted to the single-phase non-linear model of Groot *et al.* (1996), calculating the fermentation parameters A, B and C with the NLIN procedure of SAS 9.0 (SAS, 2004).

$$G = A / (1 + (B / t))^c$$

Where:

G.- Mean gas production (mL 0.2 g of DM) for a given incubation time.

A.- Gas production asintote (mL 0.2 g of DM).

B.- Time (h) after incubation at which half of the gas production has been reached.

C.- Constant that determines the shape and characteristics of the profile of the curve and, therefore, the position of the inflection point.

t.- Predictor variable representing incubation time in hours.

To analyse parameters A, B and C, the PG of the blank was subtracted from each of the samples (triplicate) to obtain the total amount accumulated in each of the hours of *in vitro* fermentation. Subsequently, the accumulated values of each repetition in each hour of sampling were analysed with the model of Groot *et al.* (1996) to obtain these parameters, these were analysed with PROC GLM to establish by Tukey the possible differences between the means of the treatments, declaring significant effect when the values of *P* were < 0.05.

RESULTS AND DISCUSSION

In vitro **digestibility of DM, NDF, FDA and LDA content of the diet.**

Digestibility of dry matter. Table 10 shows the digestibility of DM, NDF, FDA and LDA content by treatment. The highest DM digestibility was observed in T2 and T3 with values of 72.02 and 72.49 %, being higher ($P<0.05$) than T1 whose value was 64.57 %. This response has been associated with the improvement in fibre digestibility, specifically FDN and FDA shown by both treatments, as well as the lower LDA content.

Sauvant *et al.* (2004) observed a tendency to increase organic matter (OM; 0.5 %) digestibility in Holstein cows in the dry period that received a yeast culture in their diet with respect to the control group. On the other hand, the same year they published that supplementation with Sc yeast cultures (0, 2.5 and 5 g/d) in Nubian goats increased dietary OM digestibility by 12.1 and 10.1 % and NDF digestibility by 13.3 and 10.5 % for 2.5 and 5 g/d with respect to the control diet (Fadel El-seed *et al.*, 2004). In another work, it was reported that DM, NDF and FDA digestibility of bersin hay was higher when Sc yeast inoculum was added (22.5 g/d), when contrasted with the addition of 11.25 g/d in the basal diet of fattening lambs (El-Waziry and Ibrahim, 2007).

In this study, higher DM digestibility was observed in T2 and T3, showing an increase of 11.54 and 12.27 % with respect to T1, perhaps because BMZN and IL improved the microbial environment of cellulolytic bacteria and consequently increased DM digestibility.

In *vitro* digestibility of fibre fractions of diets containing fermented apple bagasse and a yeast inoculum.

Digestibility (%)	T1	Treatment[1]	T3	_EE (±)
		T2		
MS	64.57[b]	72.02[a]	72.49[a]	0.82
FDN	67.32[b]	70.77[a]	70.25[a]	0.82
FDA	60.53[b]	64.07[a]	64.56[a]	0.82
LDA	5.55[a]	4.25[b]	4.14[b]	0.04

[ab] Means with different row literals are different ($P<0.05$) between treatments. [1] T1 = Oat hay (HA) + corn silage (EM) + concentrate; T2 = HA + EM + concentrate + fermented apple bagasse (BMZN); T3 = HA + EM + concentrate + yeast inoculum (IL).

This is in agreement with other authors, who reported that yeast supplementation increases nutrient digestibility by stimulating the growth of the rumen microbial population (Harrison *et al.*, 1988). It has also been suggested that yeasts consume the oxygen available at the surface of the fresh feed intake to maintain metabolic activity and thus reduce the redox potential in the rumen (Chaucheyras-Durand *et al.*, 2008). These changes establish better conditions for the growth of strict anaerobic bacteria such as cellulolytic bacteria, stimulating their binding to forage particles and causing an increase in the initial rate of fibre degradation (Roger *et al.*, 1990), producing growth factors such as organic acids and vitamins (Chaucheyras *et al.*, 1995).

On the other hand, Ahmed and Salah (2002) developed an experiment with two levels of yeast culture (0, 4, 8 g/d) in the diet of fattening lambs, and reported that the DM digestion coefficient was improved at both levels of yeast when compared with the control diet, while the digestibility of CP was only significant between the control and the group fed 8 g/d.

It has also been mentioned that the type of diet can determine the effect of yeast on digestibility. Tang *et al.* (2008) studied the effect of yeast cultures on *in vitro* fermentation characteristics of rice, wheat and corn stubble, and reported that yeast culture (0, 2.5 and 7.5 g/kg DM) increased *in vitro* DM digestibility for each type of stubble. On the other hand, by adding a Sc yeast culture (10 g/d) to three steer diets consisting of 75 % alfalfa silage and 25 % barley, 96 % corn silage and 4.0 % soybean meal, 75 % rolled barley grain and 25 % alfalfa hay, reported that the digestibility coefficient of the diets for DM, CP, FDA and NDF did not differ with the inclusion of yeast, except for the high-grain diet, in which the yeast culture increased the digestibility of DM and CP (Mir and Mir, 1994).

Digestibility of NDF and FDF. The highest percentage in NDF digestibility was presented by T2 and T3 with values of 70.77 and 70.25 %, being higher (P<0.05) than T1 with a value of 67.32 %. Likewise, both treatments also showed the highest FDA digestibility reaching 64.07 and 64.56 % for T2 and T3 with respect to T1 with a value of 60.53 %. These results coincide with Kholif and Khorshec (2006) who reported a positive effect of yeast supplementation to lactating buffaloes on the digestibility of NDF, FDA, cellulose, hemicellulose and CP. On the other hand, other authors mentioned that the digestibility of CP and FDA was significantly increased by the addition of yeast culture of 10 and 20 g/d in fresh cows fed a basal diet of corn silage, the digestibility of CP with 0, 10 and 20 g/d showed values of 78.5, 80.8 and 79.5 % and the digestibility of FDA of 54.4, 60.2 and 56.8 %, respectively (Wohlt *et al.,* 1998).

In the present study, the BMZN treatment and the one with added IL had higher NDF and FDA digestibility. This may be because they stimulated the microbial environment by increasing the number of cellulolytic bacteria and improving fibre degradation. This coincided with Koul *et al.* (1998) who observed an increase in the number of total bacteria, cellulolytic and proteolytic bacteria when a yeast culture was added to non-lactating Holstein cows. Likewise, Harrison *et al.* (1988) commented that nutrient digestibility increases when using yeast cultures in ruminant diets, which is attributed to the stimulation of the growth in number of the rumen microbial population.

Having a direct effect with the fibrous components of the diet, being those fermented, but with slow progress from the rumen muscle to the subsequent digestive tract, have a great effect on rumen filling time (Allen, 1996). On the other hand, Fadel El-seed *et al.* (2004) published that increased NDF digestibility can decrease the rumen filling effect, which can increase feed intake.

Digestibility of LDA. The highest LDA content (P<0.05) was revealed by T1 with 5.55 % compared to T2 (4.25) and T3 (4.14). This response has been associated to the low digestibility of DM, NDF and FDA which caused the increase of LDA. Lignin is the main component of plant cell wall structure, increasing through maturity (Guo *et al.,* 2001), which affects plant tissue digestibility (Van Soest, 1994), as it is generally attached to the structural carbohydrates of cell walls providing support to the plant (Whetten and Sederoff, 1995), which is negatively correlated with the quality and digestibility of forage by ruminants (Sewalt *et al.,* 1996).

***In vitro* gas production**

The accumulated *in vitro* gas production is presented in graph 1, where it is observed that T3 showed the highest (P<0.05) volume of gas produced (1.6 mL/0.2 g of DM) from hour 6 of incubation with respect to T1. However, at hour 24, T3 was higher (P<0.05) than T1 and T2 with a gas volume of 4.20, 3.40 and 1.97 mL/0.2 g of DM, respectively. Following the same trend until hour 96. The importance of this variable states that feeds with a high gas production during the initial hours of fermentation will increase voluntary intake, resulting from a high rate of digestion (Fadel El-seed *et al.,* 2004). Degradation of a substrate will start with easily digestible fractions such as highly fermentable carbohydrates like starch. Therefore, the behaviour observed in this experiment can be explained by the lower amount of FDN, FDA and LDA in the treatments with the addition of BMZN and IL, so we can assume that they favoured microbial colonisation in both treatments and therefore improved the digestibility of the diet.

In vitro fermentation parameters

Table 11 presents the parameters of the fermentation kinetics. Parameter **(A)** represents the amount of gas production (mL), where T2 and T3 obtained the highest (P<0.05) gas production (9.11 and 9.45 mL) when compared to T1 which had 6.27 mL. This implies that T2 and T3 will have a shorter ruminal residence time resulting in a higher volume of gas production.

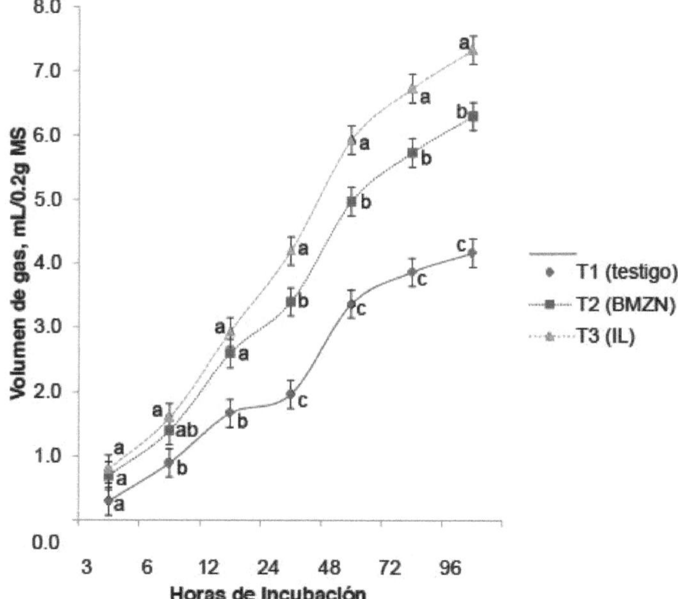

Mean (± SE) gas production of T1 (oat hay, corn silage and concentrate), T2 (oat hay, corn silage, concentrate and fermented apple bagasse) and T3 (oat hay, corn silage, concentrate and yeast inoculum) during *in vitro* ruminal fermentation.

[abc] Means with different literal in time are different (P<0.05) between treatments.

Table 11. Parameters of rumen degradability of DM from three different diets

Parameter	Treatment[1]			EE(±)
	T1	T2	T3	
A	6.27[b]	9.11[a]	9.45[a]	0.32
B	45.03[a]	39.9[a]	29.13[a]	4.48
C	0.94[a]	0.89[a]	1.02[a]	0.04

[ab] Means with different row literals are different (P<0.05) between treatments. A = Asintota in gas production (mL 0.2 g DM).

B = Time (h) at which half of the asymptotic production is reached.

C = Gas production rate (mL h).

[1] T1 = Oat hay (HA) + corn silage (EM) + concentrate; T2 = HA + EM + concentrate + fermented apple bagasse (BMZN); T3 = HA + EM + concentrate + yeast inoculum (IL).

This may be because the ingredients of the treatments in mention stimulated the microbial environment causing an increase in the rate of fermentation and numerically decreasing the time needed to reach half of the asintotic production.

The parameter **(B)**, expresses the time during incubation (h) that is required to reach half of the asymptote in gas production, which did not present significant difference (P>0.05) between treatments. T2 and T3 presented the shortest time numerically (39.9 and 29.13 h) with respect to T1 (45.03 h). This parameter explains the ruminal fermentation time due to the improvement in the chemical composition of the diet, perhaps because BMZN and IL improved the microbial environment leading to an increase in fibre digestibility and therefore, an increase in the concentration of substrates available for the microorganisms (Cone *et al.*, 1998). The biological function of this parameter indicates that diets that included BMZN and IL will require 5.13 and 15.90 h less rumen fermentation compared to T1 to achieve half the asintot in gas production, which is equal to a higher rate of ruminal digestion, allowing these diets to favour feed intake. Parameter **(C)**, indicating the rate of gas production (mL/h), showed no significant difference (P>0.05) between treatments. However, T3 presented numerically a higher gas production rate revealing 1.02 mL/h compared to T1 and T2 which presented 0.94 and 0.89 mL/h. This behaviour suggests that T3 has a higher gas production rate than T1 and T2. This behaviour suggests that T3 has a higher fermentation rate than the rest of the treatments, which is marked by the fact that this diet requires less fermentation time (parameter **B**) to reach half the asintotic gas production (parameter **A**).

AGV Production and Profiling

Table 12 shows the VFA production and profiles between treatments, where it is observed that T2 and T3 presented higher (P<0.05) concentrations of acetic (16.00 and 17.19 mmol/L), propionic (6.54 and 6.13 mmol/L) and butmic (2.63 and 2.67 mmol/L) acids when compared to T1 which had values of 12.06, 3.52 and 1.21 mmol/L, respectively.

It has been mentioned that the pattern of fermentation in ruminants takes place in the rumen environment which is influenced by the interaction between the diet, the microbial population and the animal itself (Allen and Mertens, 1988). Likewise, Rodriguez and Llamas (1990) reported that VFA production is related to methane production and fermentative balance must be maintained at all times, because methane and propionate serve as scavengers for excess reducing equivalents produced at the rumen level. The addition of yeast cultures in ruminant feeds has shown contradictory effects on rumen VFA concentration. Corona *et al.* (1999) reported that steers and lambs supplemented with 7.5 and 3 g/d of a Sc yeast culture showed low concentrations of total VFA and butmic acid molar ratio, respectively. On the other hand, they published that total rumen VFA concentration and the ratio of acetic, propionic and butmic acids were not affected by the addition of yeast cultures (Pinos-Rodriguez *et al.*, 2008). However, Harrison *et al.* (1988) found a decrease in the molar ratio of acetic acid and an increase in the molar concentration of propionic acid in the rumen fluid of Holstein cows supplemented with 114 g/d of a yeast culture containing Sc.

Table 12. Behaviour in VFA production and profiles between treatments during *in vitro* fermentation.

Variable	Treatments[1]			EE(±)
	T1	T2	T3	
Acetic acid (mmol/L)	12.06[b]	16.00[a]	17.19[a]	0.90
Propionic acid (mmol/L)	3.52[b]	6.54[a]	6.13[a]	0.75
Butamic acid (mmol/L)	1.21[b]	2.63[a]	2.67[a]	0.41

[ab] Means with different row literals are different (P<0.05) between treatments.

[1] T1 = Oat hay (HA) + corn silage (EM) + concentrate; T2 = HA + EM + concentrate + fermented apple bagasse (BMZN); T3 = HA + EM + concentrate + yeast inoculum (IL).

In the present investigation, an increase in the concentration of acetic, propionic and butynic acids was observed in the diet with the addition of BMZN and an IL, probably due to the improved rumen environment and consequently the increased number of cellulolytic bacteria leading to a higher digestibility of DM, NDF and FDA as shown in Table 10.

Koul *et al.* (1998) reported an increase in total VFA in the rumen of buffalo calves fed 5 g/d of a Sc-containing yeast culture when compared to the control group (132.2 and 122.4 mmol/L). In another study, Dolezal *et al.* (2005) observed an increase in VFA production by increasing the dose of the Sc yeast culture (strain SC-47) in the feed of lactating Holstein cows.

Concentration of N-NH3, Lactic Acid and pH

Table 13 presents the concentration of N-NH3, lactic acid and pH between treatments. The highest (P<0.05) N-NH3 concentration was shown by T2 and T3 revealing values of 0.21 and 0.22 mM/mL when compared to T1. Likewise, both treatments were different (P<0.05) in lactic acid concentration (1.46 and 1.37 mM/mL) with respect to T1. On the other hand, a higher pH (6.74) (P<0.05) was observed in T3 compared to T1 and T2. The latter maintained a higher pH (6.56) (P<0.05) compared to T1.

Ammonia nitrogen (N-NH3). Several sources of nitrogen contribute to ammonia production, such as non-protein nitrogen (NNP) from the diet, salivary nitrogen and possibly small amounts of urea that penetrate through the rumen epithelium (Moloney and Drennan, 1994).

Table 13. Means (± EE) of N-NH3 concentration, lactic acid and pH behaviour of treatments during *in vitro* fermentation.

| Variable | T1 | Treatments[III][IV] | | T3 | EE(±) |
		T2			
N-NH3 (mM/mL)	0.18[b]	0.21[a]		0.22[a]	0.01
Lactic acid (mM/mL)	2.30[a]	1.46[b]		1.37[b]	0.20
pH	6.19[c]	6.56[b]		6.74[a]	0.06

In the present study, an increase in N-NH3 concentration (0.21 and 0.22 mM/mL) was observed in the diets with the addition of BMZN and an IL, perhaps due to the improved rumen environment and consequently increased digestibility of the diet, which coincides with Newbold *et al.* (1995) where they observed an increase in N-NH3 concentration in rumen fermentation *in vitro*, attributing the result to increased availability of substrate for microorganisms. Low pH inhibits N-NH3 production *in vitro* because it affects methanogenic bacteria (*Methanobacterium bryantii, M. formicicum and Methanosarcina barkeri*) and protozoa in fibre degradation (Nagaraja and Titgemeyer, 2007).

Lactic acid and pH. Williams *et al.* (1983) observed that low rumen pH reduces the viability of cellulolytic bacteria (*Ruminococcus albus and Fibrobacter succinogenes*) and thus reduces activity on structural carbohydrates. In another study it was concluded that in low pH rumen conditions, bacterial attack on cell walls is reduced and consequently their digestion is reduced (Cheng *et al.*, 1984), so it is considered that a rumen pH above 6.2 is the optimum for good cellulose digestion (Rodriguez and Llamas, 1990).

In this research, it was observed that the treatments with the addition of BMZN and IL decreased lactic acid concentration, while T3 increased pH, perhaps because they stimulated lactate-consuming bacteria (Megasphaera elsdenii and Selenomonas ruminantium) and therefore the IL stimulated the microbial environment to maintain an optimal pH, This is consistent with Robinson (2010), who observed that modulation of rumen pH is one of the effects of the addition of yeast in the diet, exerting an average increase in rumen pH, exerting an average increase in rumen pH (1.6 %), an overall increase in total VFA (5.4 %) and an overall decrease in lactate concentration (8.1 %).

[abc] Means with different row literals are different (P<0.05) between treatments.

[IV] T1 = Oat hay (HA) + corn silage (EM) + concentrate; T2 = HA + EM + concentrate + fermented apple bagasse (BMZN); T3 = HA + EM + concentrate + yeast inoculum (IL).

(2004) observed no effect of yeast culture on VFA concentration and rumen pH, revealing only a tendency to increase DM digestibility (+0.5 %) in Holstein cows in the dry period. The meta-analysis study by Desnoyers *et al.* (2009) suggested that yeast supplementation increased VFA concentration (2.1 mmol/L) and ruminal pH, along with a tendency to decrease lactate concentration in Holstein cows in transition. The same authors reported that yeasts are able to limit the reduction of rumen pH which is usually related to an increase in VFA concentration. On the other hand, yeasts are able to limit the production of lactic acid or its accumulation in the rumen (Desnoyers *et al.*, 2009), probably because Sc can compete with starch fermenting bacteria (Lynch and Martin, 2002), preventing lactate accumulation in the rumen (Chaucheyras *et al.*, 1995).

CONCLUSIONS AND RECOMMENDATIONS

The addition of BMZN and an IL in the diet of weaned calves improved DM, NDF and FDA digestibility, although it was not determined whether these improvements would promote better performance of the animals. Also, the addition of these ingredients resulted in increased gas and VFA production due to increased digestibility compared to the control diet. Also, the addition of fermented apple bagasse and yeast inoculum in the diet increased the concentration of N-NH3 and pH, decreasing the concentration of lactic acid.

Based on the results obtained in this study, it is recommended to develop further research with BMZN supplements and an IL, considering the priority of maximising fibre digestion, it is necessary to evaluate these supplements under different levels of inclusion in order to achieve a better animal response.

LITERATURE CITED

Ahmed, B. M. and M. S. Salah. 2002. Effect of yeast culture as an additive to sheep feed on performance, digestibility, nitrogen balance and rumen fermentation. J. King Saud. Univ. Agric. Sci. 14:1-13.

Akin, D. E. 1989. Histological and physical factors affecting digestibility of forages. Agron. J. 81: 17-25.

Allen, M. S. 1996. Physical constraints on voluntary intake of forages by ruminants. J. Anim. Sci. 74:3063-3075.

Allen, M. S. and M. Mertens. 1988. Evaluating constraints on fiber digestion by rumen microbes. J. Nutr. 118:261-270.

AOAC. 2000. Official Methods of Analysis. Vol. I. 16th ed. International, Arlington, VA. U.S.A.

Beauchemin, K. A., D. Colombatto, D. P. Morgavi and W. Z. Yang. 2003. Use of exogenous fibrolytic enzymes to improve feed utilization by ruminants. J. Anim. Sci. 81:37-47.

Blummel, M., J. W. Cone, A. H. Van Gelder, I. Nshalai, N. N. Umunna, H. P. S. Makkar and K. Becker. 2005. Prediction of forage intake using *in vitro* gas production methods: Comparison of multiphase fermentation kinetics measured in an automated gas test, and combined gas volume and substrate degradability measurements in a manual syringe system. Anim. Feed Sci. Technol. 123:517-526.

Broderick, G. A. and J. H. Kang. 1980. Automated simultaneous determination of ammonia and total amino acids in ruminal fluid and *in vitro* media. J. Dairy. Sci. 63:64-75.

Brotz, P. G. and D. M. Schaeffer. 1987. Simultaneous determination of lactic acid and volatile fatty acids in microbial fermentation extracts by gas-liquid chromatography. J. Microbiol. Methods. 6:139-144.

Bruni M. de los A. and P. Chilibroste. 2001. Simulation of ruminal digestion by gas production method. Arch. Latinoam. Prod. Anim. 9: 43-51.

Chaucheyras, F., G. Fonty, G. Bertin and P. Gouet. 1995. *In vitro* H2 utilization by a ruminal acetogenic bacterium cultivated alone or in association with an *Archaea methanogen* is stimulated by a probiotic strain of *Saccharomyces cereviciae*. Appl. Environ. Microbiol. 61:3466-3467.

Chaucheyras-Durand, F., N. D. Walker and A. Bach. 2008. Effects of active dry yeasts on the rumen microbial ecosystem: Past, present and future. Anim. Feed Sci. Technol. 145:5-26.

Cheng, K. J., C. S. Stewart, D. Dinsdale, and J. W. Costerton. 1984. Electron microscopy of the bacteria involved in the digestion of plant cell walls. Anim. Feed Sci. Technol. 10:93-101.

Cone, J. W., A. H. Van Gelder and H. Valk. 1998. Prediction of nylon bag degradation characteristics of grass samples with the gas production technique. J. Sci. Food Agric. 77:421-426.

Corona, L., G. D. Mendoza, F. A. Castrejon, M. M. Crosby and M. A. Cobos. 1999. Evaluation of two yeast cultures (*Saccharomyces cerevisiae*) on ruminal fermentation and digestion in sheep fed a corn stover diet. Small Rumin. Res. 31:209-214.

Desnoyers, M., S. Giger-Reverdin, G. Bertin, C. Duvaux-Ponter and D. Sauvant. 2009. Meta-analysis of the influence of *Saccharomyces cerevisiae* supplementation on ruminal parameters and milk production of ruminants. J. Dairy Sci. 92:1620-1632.

Dolezal, P., J. Dolezal and J. Trinacty. 2005. The effect of *Saccharomyces cerevisiae* on ruminal fermentation in dairy cows. Czech J. Anim. Sci. 50:503-510.

El-Waziry, A. M. and H. R. Ibrahim.2007. Effect of *Saccharomyces cerevisiae* of yeast on fiber digestion in sheep fed berseem (*Trifolium alexandrinum*) hay and cellulase activity. Aust. J. Basic. Applied Sci. 1:379-385.

Fadel El-seed, A. N. M. A., J. Sekine. H. E. M. Kamel and M. Hishinuma. 2004. Changes with time after feeding in ruminal pool sizes of cellular contents, crude protein, cellulose, hemicellulose and lignin. Indian J. Anim. Sci. 74:205-210.

Goering, H. K. and P. J. Van Soest. 1970. Forage Fiber Analyses (apparatus, reagents, procedures and some applications). P. 379 in Agric. Handbook. USDA-ARS, Washington, DC. U. S. A.

Groot, J. C. J. J., J. W. Cone, B. A. Williams, F. M. A. Debersaques and E. A. Lantinga. 1996. Multiphasic analysis of gas production kinetics for *in vitro* fermentation of ruminant feeds. Anim. Feed Sci. Technol. 64:77-89.

Guo, D., F. Chen, K. Inoue, J. W. Blount and R. A. Dixon. 2001. Downregulation of caffeic acid *3-O-methyltranferase* and caffeoyl CoA *3-O- methyltranferase* in transgenic alfalfa: Impacts on lignin structure and implications for the biosynthesis of G and S lignin. Plant Cell. 13:73-88.

Harrison, G. A., R. W. Hemken, K. A. Dawson, R. J. Harmon and K. B. Barker. 1988. Influence of addition of yeast culture supplement to diets of lactating cows on ruminal fermentation and microbial population. J. Dairy Sci. 71:2967-2975.

Hoover, W. H. 1986. Chemical factors involved in ruminal fibre digestion. J. Anim. Sci. 69:2755-2768.

INAFED, 2008. National Institute for Federalism and Municipal Development. http://www.inafed.gob.mx/work/templates/enciclo/chihuahua/. Accessed September 10, 2012.

Kholif, S. M. and M. M. Khorshed. 2006. Effect of yeast or selenized yeast supplementation to rations on the productive performance of lactating buffaloes. Egypt. J. Nutr. Feed. 9:193-205.

Koul, V., U. Kumar, V. K. Sareen and S. Singh. 1998. Mode of action of yeast culture (Yea-Sacc[1026]) for stimulation of rumen fermentation in buffalo calves. J. Sci. Food Agric. 77:407-413.

Lesmeister, K. E., A. J. Heinrichs and M. T. Gabler. 2004. Effects of supplemental yeast (*Saccharomyces cerevisiae*) culture on rumen development, growth characteristics, and blood parameters in neonatal dairy calves. J. Dairy Sci. 87:1832-1839.

Lynch, H. A. and S. A. Martin. 2002. Effects of *Saccharomyces cerevisiae* culture and *Saccharomyces cerevisiae* live cells on in vitro mixed ruminal microorganism fermentation. J. Dairy Sci. 85:2603-2608.

Menke, K. H., L. Raab, A. Salewski, H. Steingass, D. Fritz and W. Schneider. 1979. The estimation of the digestibility and metabolizable energy content of ruminant feedstuffs from the gas production when they are incubated with rumen liquor *in vitro*. J. Agric. Sci. Camb. 93:217-222.

Menke, K. H. and H. Steingass. 1988. Estimation of the energetic feed value obtained from chemical analysis and *in vitro* gas production using rumen fluid. Anim. Research Develop. 28:7-55.

Mir, Z. and P. S. Mir. 1994. Effect of the addition of live yeast (*Saccharomyces cerevisiae*) on growth and carcass quality of steers fed high forage or high grain diets and on feed digestibility and *in situ* degradability. J. Anim. Sci. 72:537-545.

Moloney, A. P. and M. J. Drennan. 1994. The influence of the basal diet on the effects of yeast culture on ruminal fermentation and digestibility in steers. Anim. Feed Sci. Technol. 50:55-73.

Muro, R. A. 2007. Rumen degradation kinetics of three forage sources using *in vitro* digestibility by gas production. Doctoral dissertation. Faculty of Animal Science. Autonomous University of Chihuahua. Chihuahua. Chih. Mex.

Nagaraja, T. G. and E. C. Titgemeyer. 2007. Ruminal acidosis in beef cattle: The current microbiological and nutritional outlook. J. Dairy Sci. 90:17-38.

Newbold, C. J., R. J. Wallace, X. B. Chen, and F. M. Mcintosh. 1995. Different strains of

Saccharomyces cerevisiae differ in their effects on ruminal bacterial numbers *in vitro* and in sheep. J. Anim. Sci. 73:1811-1818.

Olson, K. C., J. S. Caton, D. R. Kirby and P. L. Norton. 1994. Influence of yeast culture supplementation and advancing season on steers grazing mixed- grass prairie in the northern great plains: II. Ruminal fermentation site of digestion, and microbial efficiency. J. Anim. Sci. 72:2158-2170.

Pinos-Rodriguez, J. M., P. H. Robinson, M. E. Ortega, S. L. Berry, G. Mendozad and R. Barcena. 2008. Performance and rumen fermentation of dairy calves supplemented with *Saccharomyces cerevisiae 1077* or *Saccharomyces boulardii 1079*. Anim. Feed Sci. Technol. 140:223-232.

Plata, P. F., M. G. D. Mendoza, J. R. Barcena-Gama and M. S. Gonzalez. 1994. Effect of a yeast culture (*Saccharomyces cerevisiae*) on neutral detergent fiber digestion in steers fed oat straw based diets. Anim. Feed Sci. Technol. 49:203-210.

Robinson, P. H. 2010. Yeast products for growing and lactating ruminants: A literature summary of impacts on rumen fermentation and performance. http://animalscience.ucdavis.edu/faculty/robinson/Articles/FullText/pdf/W e b200901.pdf. Accessed October 16, 2012.

Rodriguez, G. F. and L. G. Llamas. 1990. Digestibility, Nutrient Balance and Rumen Fermentation Patterns. In: R. A. Castellanos, L. G. Llamas and S. A. Shimada, Ed. Manual de tecnicas de investigation en ruminolog^a. Continuing education systems in animal production in Mexico. A. C. Mexico, D. F. Mexico.

Roger, V., G. Fonty, S. Komisarczuk-Bony and P. Gouet. 1990. Effects of physicochemical factor on the adhesion to cellulose, avicel of the rumen bacteria *Ruminococcus flavefaciens* and *Fibrobacter succinogenes*. Applied Environ. Microbiol. 56:3081-3087.

SAS. 2004. SAS/STAT®9.1 User's Guide. SAS Institute Inc. United States of America.

Sauvant, D., S. Giger-Reverdin and P. Schmidely. 2004. Rumen Acidosis: Modeling Ruminant. Proceeding of the alltech's 20th annual symposium: Reimagining the feed industry, May 23-26, 2004. Nottingham University Press. London. London. United Kingdom.

Steel, R. G. D. and J. H. Torrie. 1997. Biostatistics. Principles and Procedures. 2nd ed. McGraw-Hill. Mexico.

Sewalt, V. J. H., W. G. Glasser, J. P. Fontenot and V. G. Allen. G. Allen. 1996. Lignin impact on fiber degradation. 1 quinone methide intermediates formed from lignin during *in vitro* fermentation of corn stover. J. Sci. Food Agric. 71:195-203.

Tang, S. X., G. O. Tayo, Z. L. Tan, Z. H. Sun, and L. X. Shen. 2008. Effects of yeast culture and fibrolytic enzyme supplementation on *in vitro* fermentation characteristics of low quality cereal straws. J. Anim. Sci. 86:1164-1172.

Taylor, K. A. C. C. C. 1996. A simple colorimetric assay for muramic acid and lactic acid. Appl. Biochemist and Biotechnol. 56:49-58.

Theodorou, M. K., B. A. Williams, M. S. Dhanoa, A. B. McAllan and J. France. 1994. A simple gas production method using a pressure transducer to determine the fermentation kinetics of ruminant feeds. Anim. Feed. Sci. Technol. 48:185-197.

Van Soest, P. J., J. B. Robertson, and B. A. Lewis. 1991. Methods for dietary fiber, neutral detergent fiber, and non-starch polysaccharides in relation to animal nutrition. J. Dairy Sci. 74:3579-3583.

Van Soest, P.J. 1994. Nutritional Ecology of the Ruminant. 2nd ed. Cornell University Press. Ithaca, New York. U.S.A.

Whetten, R. and R. Sederoff. 1995. Lignin biosynthesis. The plant cell. 7:10011013.

Williams, P. E. V., A. Macdearmid, G. M. Innes and A. Brewer. 1983. Turnips with chemically treated straw for beef production. Effect of turnips on the degradability of straw in the rumen. Anim. Prod. 37:189-196.

Williams, P. E. V., C. A. G. Tait, G. M. Innes, and C. J. Newbold. 1991. Effects of the inclusion of yeast culture (*Saccharomyces cerevisiae* plus growth medium) in the diet of cows on milk yield and forage degradation and fermentation patterns in the rumen of sheep and steers. J. Anim. Sci. 69:3016-3026.

Wohlt, J. E., T. T. Corcione and P. K. Zajac. 1998. Effect of yeast on feed intake and performance of cows fed diets based on corn silage during early lactation. J. Dairy Sci. 81:1345-1352.

IN VITRO FERMENTATION OF DIETS FOR GROWING CALVES
SUPPLEMENTED WITH FOUR STRAINS OF
YEAST

SUMMARY

The objective was to evaluate the effect of *in vitro* fermentation of growing calf diets supplemented with four yeast strains. Treatments consisted of: T1 (Sc6): oat hay (HA) + corn silage (EM) + concentrate + Sc6 strain; T2 (Kl2): HA + EM + concentrate + Kl2 strain; T3 (Kl11): HA + EM + concentrate + Kl11 strain and T4 (Io3): HA + EM + concentrate + Io3 strain. The variables measured were dry matter (DM) digestibility, neutral detergent fibre (NDF), acid detergent fibre (ADF), and acid detergent lignin (ADL) content, gas production volume, concentration of acetic acid, propionic acid, butmco, ammoniacal nitrogen ($N\text{-}NH_3$), lactic acid and pH. *In vitro* digestibility of the diets was performed at 48 h, while for the rest of the variables, samples were taken at 3, 6, 12, 24, 48, 72 and 96 h. A completely randomised design was used for fibre digestibility, using a model that included a fixed treatment effect as a fixed effect. For volatile fatty acids (VFA), $N\text{-}NH_3$, lactic acid and pH, treatment and time were included as fixed effects. Gas production was analysed with a similar model that included treatment, time and their interaction as fixed effects. To analyse parameters A, B and C of the fitted non-linear regression model, treatment was used as a fixed effect. Treatment effect ($P<0.05$) was observed in DM digestibility, being higher in T2, T3 and T4 with values of 63.01, 63.11 and 63.05 %, with respect to T1 (61.30 %). Likewise, T4 was higher ($P<0.05$) in NDF digestibility (56.69 %) when compared to T1 and T2 (48.22 and 53.98 %), T2 and T3 had better ($P<0.05$) FDA digestibility (58.78 and 58.94 %) when compared to T1 (56.39 %), therefore, the lowest ($P<0.05$) LDA content was for T2, T3 and T4 (4.54, 4.42 and 4.46). At 96 h, T3 showed the highest ($P<0.05$) volume of gas production (4.54, 4.42 and 4.46).
(11.50 mL/0.2 g DM). The highest concentration ($P<0.05$) of VFA was presented by T2 with values of 37.30 and 19.25 mmol/L for acetic and propionic acid. Also, T1, T2 and T4 were higher ($P<0.05$) in butamic acid concentration (7.45, 7.74 and 6.73 mmol/L). T4 showed an increase ($P<0.05$) in $N\text{-}NH_3$ and lactic acid concentration (5.34 and 0.69 mM/mL), with T1 showing an increase ($P<0.05$) in pH (7.00). It is concluded that strains Kl2, Kl11 and Io3 favoured DM, NDF and FDA digestibility, with respect to Sc6, and Kl2 increased the concentration of acetic and propionic acids during *in vitro* fermentation of diets for growing steers.

INTRODUCTION

The use of yeast strains as an additive in fibrous diets for ruminants produces improvements in the efficiency of nutrient utilisation and availability and increases in the rumen digestion of dry matter (DM), organic matter (OM), neutral detergent fibre (NDF), acid detergent fibre (ADF) and nitrogen, both *in vivo* and *in vitro* (Biricik and Turkman, 2001). However, the mechanisms by which yeasts exert their action in the rumen have not been properly elucidated. The different modes of action of yeast strains and the interaction with the diet offered to the animals present new opportunities as well as new problems in defining the modification of the rumen metabolism caused by the effect of the different culture strains and the amount fed (Karma *et al.*, 2002). Brossard *et al.* (2006) reported that yeasts improve the rumen environment by causing an increase in the population of cellulolytic bacteria, and stimulate the growth of lactate consuming bacteria (*Selenomonas ruminantium and Megasphaera elsdenii*) which results in an increase in fibre degradation (Callaway and Martin, 1997) and changes in volatile fatty acids (Carro *et al.*, 1992). On the other hand, the *in vitro* gas production technique simulates the digestive processes generated by microbial production, allowing to know the fermentation and degradability of the feed as a function of the nutritional quality and availability of nutrients for the bacteria (Theodorou *et al.*, 1994).

The objective of the study was to evaluate the effect of four yeast strains added to diets of weaned calves on *in vitro* fibre digestibility, gas production and volatile fatty acid (VFA) profile.

MATERIALS AND METHODS

Location of the Study Area

This research was carried out in the facilities of the Facultad de Zootecnia y Ecolog^a of the Universidad Autonoma de Chihuahua, Chih., Mexico, located in the coordinates 28° 35' 07" north latitude and 106° 06' 23" west longitude, with an altitude of 1,517 masl, according to the Global Positioning System (GPS, MAGELLAN® , MobileMapper-Pro). The average annual temperature is 18.2 °C, with an average maximum of 37.7 °C and an average minimum of -7.4 °C. The average annual precipitation is 387.5 mm, with 71 rainy days per year and a relative humidity of 49 %; predominantly an extreme semiarid climate (INAFED, 2008).

Description of Treatments

For the development of the present experiment, the diet of treatment T1 (control) was used: oat hay (HA) + corn silage (EM) + concentrate, offered to the weaned calves as mentioned in Experiment I. The experimental treatments were four inocula prepared with the following strains: T1: *Saccharomyces cerevisiae* strain 6, T2: *Kluyveromyces lactis* strain 2, T3: *Kluyveromyces lactis* strain 11 and T4: *Issatchenkya orientalis* strain 3, for which 4 Erlenmeyer flasks (Kimax)® of 1,000 mL were used, adding 3.1×10^8 live yeast cells of each strain per flask, likewise, the ingredients shown in Table 14 were added to the culture. Once the fermentation time was over, the yeast count was carried out following the procedure of D^az (2006) as indicated in Experiment I, finding the amount of 3.4×10^9 cells/mL in order to subsequently adjust the amount of yeast cells per treatment,

Table 14. Treatment design for the preparation of the inocula with the four yeast strains

Ingredients	Treatments[1]			
	T1	T2	T3	T4
Yeast strain	Sc6	Kl2	Kl11	Io3
Cane molasses (g)	100	100	100	100
Urea (g)	1.2	1.2	1.2	1.2
Minerals (g)	0.5	0.5	0.5	0.5
Ammonium sulphate (g)	0.2	0.2	0.2	0.2
Aphorates (mL)	1000	1000	1000	1000

[1]T1 = *Saccharomyces cerevisiae* strain 6; T2 = *Kluyveromyces lactis* strain 2; T3 = *Kluyveromyces lactis* strain 11 and T4 = *Issatchenkya orientalis* strain 3.

based on 26 mL added to the daily diet received by each animal.

Chemical analysis of diets

In order to obtain the composite values of dry matter (DM) digestibility, neutral detergent fibre (NDF), acid detergent fibre (ADF) and acid detergent lignin (ADL) content, the chemical composition of the diet fractions with the addition of the above mentioned strains was analysed. Diet samples were collected every 15 d to form composite samples per month, using 1 kg DM (complete ration) per treatment. Samples were dried at 60 °C for 48 h in a forced-air oven and ground to 1 miKmeter (mm) in a Wiley mill (Arthur H. Thomas Co., Philadelphia, PA), these samples were dried at 105 °C for 8 h in a forced-air oven to determine absolute DM and sequentially incinerated at 600 °C for 4 h in a muffle to determine ash (AOAC, 2000). In the composite samples, NDF, FDA (Van Soest *et al.,* 1991) and LDA (Goering and Van Soest, 1970) were determined. For NDF analysis, sodium sulphite (Na_2SO_3) and thermostable a-amylase (Ankom Technology) were used to remove nitrogenous matter and starch.

***In vitro* digestibility of DM, NDF, FDA and LDA content of the diet.**

The complete ration (1 kg of DM) ground from each treatment was added to the adjusted amount of yeast inoculum by hand mixing in a plastic tray. Prior to incubation, the bags were dipped in acetone to remove the surfactant that inhibits microbial digestion, and then dried at room temperature. Next, the composite samples of the diets (0.45 g ± 0.05) were weighed into ANKOM filter bags® F57 (25 pm pore size and dimensions of 5 x 4 cm) identified and heat sealed and incubated *in vitro* according to the Daisy methodology[11] (ANKOM Technology Corp., Fairport, NY- USA), which involves buffer solutions A, B and ruminal inoculum as indicated in Experiment II.

Rumen Fluid Collection

The same procedure as in Experiment II was followed to obtain the ruminal Kquido.

***In vitro* gas production**

In vitro gas production (PG) was performed using the protocol of Menke and Steingass (1988), with the modifications mentioned by Muro (2007). Incubation of the samples was performed in triplicate and a blank was added at each of the sampling times. A total of 112 50 mL glass vials, sealed with rubber stoppers, were used; 0.2 g of sample, 10 mL of ruminal Kquido and 20 mL of artificial saliva were added following the same procedure as in Experiment II.

The internal pressure of the jars due to feed degradation was measured with a pressure transducer (FESTO®) at 3, 6, 12, 24, 48, 72 and 96 h of incubation, by puncturing the jars and recording the accumulated pressure (Theodorou *et al.,* 1994). The results were expressed in mL PG per 0.2 g DM (mL PG/0.2 g DM).

AGV Production and Profiling

Samples of 20 mL were collected from the PG flasks to determine the production of acetic, propionic and butamic acid at 3, 6, 12, 24, 48, 72 and 96 h, and filtered with two layers of gauze to separate the solid and kieselguhr contents. The pH of the latter was immediately determined using a potentiometer (Combo, HANNA instruments® Inc., Woonsocket, RI). A 10 mL subsample was taken and centrifuged at 3,500 *xg at* 4 °C for 10 min. The supernatant was then placed in pre-labelled amber vials, acidified with 0.2 mL of 50 % H2SO4 and frozen at - 20 °C until analysis. Prior to measurement, the samples were thawed in refrigeration at 4 °C, added 25 % metaphosphoric acid and centrifuged again with the above mentioned characteristics and preserved in refrigeration until analysis by gas chromatography (Brotz and Schaefer, 1987), following the same procedure as in Experiment II.

Ammoniacal Nitrogen (N-NH3)

A 20 mL sample was obtained from the PG flasks to analyse N- NH3 at 3, 6, 12, 24, 48, 72 and 96 h by colorimetry following the technique of Broderick and Kang (1980) as mentioned in Experiment II.

Lactic acid

A 20 mL sample was collected from the PG bottles for analysis of lactic acid content at 3, 6, 12, 24, 48, 72 and 96 h by colorimetry according to Taylor (1996) following the same procedure as described in Experiment II.

Statistical Analysis

For the analysis of DM, NDF, FDA and LDA digestibility, a statistical model was fitted including treatment as a fixed effect in a completely randomised design. On the other hand, for N-NH3, acetic, propionic, butynic and lactic acid concentrations, as well as for pH a similar model was fitted, considering treatment and sampling time as fixed effects. For gas production the statistical model included the fixed effects of treatment and time, as well as their interaction. To analyse parameters A, B and C of the fitted non-linear regression model, treatment was used as a fixed effect. Tukey's test was used to establish possible differences between treatment means (Steel and Torrie, 1997).

The *in vitro* accumulated PG amounts were fitted to the non-linear Gompertz model (Lavrencic *et al.,* 1997), calculating the fermentation parameters A, B and C; with the NLIN procedure of SAS (SAS, 2004).

Y = A * (exp (- B * (exp (- C *

t))) Where:

Y.- Volume of gas production (mL 0.2 g of DM) for a given incubation time.

A.- Gas volume corresponding to the complete digestion of the substrate (time at turning point).

B.- Constant rate of gas production (mL of gas at the point of

inflexion). C.- Maximum gas production rate (mL/h).

t.- Incubation time (h).

To analyse parameters A, B and C, the PG of the blank was subtracted from each of the samples (triplicate) to obtain the total amount accumulated in each of the hours of *in vitro* fermentation. Subsequently, the accumulated values of each repetition in each hour of sampling were analysed with the model of Gompertz *et al.* (1996) to obtain these parameters, these were analysed with PROC GLM to establish by Tukey the possible differences between the means of the treatments, declaring significant effect when the *P* values were < 0.05.

RESULTS AND DISCUSSION

***In vitro* digestibility of DM, NDF, FDA and LDA content of the diet.**

Digestibility of dry matter. Table 15 shows the digestibility of DM, NDF, FDA and LDA content by treatment. The highest DM digestibility was observed in T2, T3 and T4 (63.01, 63.11 and 63.05 %, respectively) being higher (P<0.05) than T1 (61.30 %). This tendency has been associated with the increase in fibre digestibility, mainly NDF and FDA shown by the three treatments, as well as the lower LDA content.

Sauvant *et al.* (2004) observed a tendency to increase organic matter (OM; 0.5 %) digestibility in Holstein cows in the dry period that received a yeast culture in their diet with respect to the control group. On the other hand, it has been suggested that live yeasts are metabolically active in the rumen, thereby modifying fermentation and stimulating microbial growth (Erasmus *et al.*, 2005). Such changes are associated with an increase in fibre digestibility, which may increase the rate of passage and thus improve DM intake and animal productivity (Guedes *et al.*, 2008). There is little or no information on the addition of yeast strains *Kluyveromyces lactis* and *Issatchenkya orientalis* in ruminant diets. In the present work, the best performance in DM digestibility was observed in T2, T3 and T4, so these results suggest that strains KI2, KI11 and Io3 favoured the microbial environment of cellulolytic bacteria and as a consequence increased fibre digestibility. It is also suggested that these strains increased nutrient digestibility because they stimulate the growth of the rumen microbial population as indicated by Harrison *et al* (1988).

Table 15. *In vitro* digestibility of fibre fractions between treatments

Digestibility (%)	Treatments[1]				
	T1	T2	T3	T4	. EE(±)
MS	61.30[b]	63.01[a]	63.11[a]	63.05[a]	0.51
FDN	48.22[c]	53.98[b]	55.80[ab]	56.69[a]	0.67
FDA	56.39[b]	58.78[a]	58.94[a]	57.83[ab]	0.57
LDA	5.11[a]	4.54[b]	4.42[b]	4.46[b]	0.05

[abc] Means with different row literals are different (P<0.05) between treatments. [1]-T1 = Oat hay (HA) + corn silage (MS) + concentrate + *Saccharomyces cerevisiae* strain 6; T2 = HA + MS + concentrate + *Kluyveromyces lactis* strain 2; T3 = HA + MS + concentrate + *Kluyveromyces lactis* strain 11 and T4 = HA + MS + concentrate + *Issatchenkya orientalis* strain 3.

as indicated by Harrison *et al.* (1988). Chaucheyras-Durand *et al.* (2008) mentioned that yeasts consume oxygen available on the surface of the fresh feed intake to maintain metabolic activity and thus reduce the redox potential in the rumen. They also published that these changes establish better conditions for the growth of strict anaerobic bacteria such as cellulolytic bacteria, favouring their binding to feed particles and thus increasing the rate of fibre degradation (Roger *et al.*, 1990).

Digestibility of NDF and FDF. The highest percentage (P<0.05) in NDF digestibility was presented by T4 revealing 56.69 %, being higher than T1 and T2 which showed 48.22 and 53.98 %, respectively. However, the latter was superior (P<0.05) to T1. On the other hand, T2 and T3 showed higher (P<0.05) FDA digestibility (58.78 and 58.94 %) compared to T1 whose value was 56.39 %, while the value of T4 was 57.83 %.

Kholif and Khorshed (2006) published a positive effect of yeast supplementation to lactating buffaloes on the digestibility of NDF, FDA, cellulose, hemicellulose and crude protein (CP). In another investigation, they mentioned that the addition of yeast cultures in diets of fattening sheep increased the number of cellulolytic bacteria such as *Fibrobacter succinogenes* (*F.succinogenes*), *Ruminococcus albus* (*R. albus*) and *Ruminococcus flavefaciens* (*R. flavefaciens*) in the rumen, which favours cellulose degradation (Newbold *et al.*, 1995).

In the present experiment, T4 presented the highest NDF digestibility (56.69 %), while T2 and T3 showed the highest percentage of FDA digestibility (58.78 and 58.94 %), being T1 lower in NDF and FDA (48.22 and 56.39 %) with respect to the other treatments.

This behaviour suggests that strains Io3, KI2 and KI11 were able to stimulate the microbial environment by increasing the number of cellulolytic bacteria improving fibre degradation relative to Sc. This is in agreement with Koul *et al.* (1998) who observed an increase in the total number of cellulolytic and proteolytic bacteria when a yeast culture was added to nonlactating Holstein cows. Likewise, Williams *et al.* (1991) reported that stimulation of cellulose degradation by yeast culture is associated with a decrease in lag time, which results in an increase in the initial rate of digestion, but not an increase in total digestion by rumen microorganisms.

Chaucheyras-Duran and Fonty (2001) found that digesta-attaching microorganisms represent more than 75 % of the total microflora in the rumen, with *F. succinogenes, R. albus* and *R. flavefaciens* being the most active bacteria in fibre degradation due to the presence of numerous cellulases (glucanases, glucosidases) and hemicellulases (xylanases, xylosidases and fucosidases).

Digestibility of LDA. The highest LDA content (P<0.05) was found in T1 with 5.11 % compared to T2, T3 and T4, which showed 4.54, 4.42 and 4.46 %, the latter being similar to each other. The increase in LDA content has been associated with the low digestibility of DM, NDF and FDA. Noguera *et al.* (2004) mentioned that in the first hours of fermentation a part of the substrate, mainly soluble sugars are rapidly fermented, however they only constitute a small part of the potentially digestible material, and as the fermentation process continues a smaller amount of material is hydrated and colonised by rumen microorganisms, which causes different degradation rates depending on the concentration of structural carbohydrates, lignin content and plant maturity stage. Lignin is the main component affecting the digestibility of plant tissue (Van Soest, 1994), as it is generally attached to the structural carbohydrates of the cell walls providing support to the plant (Whetten and Sederoff, 1995), which is negatively correlated with the quality and digestibility of forage by ruminants (Sewalt *et al.,* 1996).

In vitro gas production

The accumulated *in vitro* gas production is presented in graph 2, where it is observed that T3 and T4 showed the highest (P<0.05) volume of gas produced (5.77 and 5.50 mL/0.2 g of DM) from 24 h of incubation with respect to T1 and T2 (3.53 and 3.40 mL/0.2 g of DM), following the same trend until 72 h. At 96 h, T3 was superior (P<0.05) with values of 11.50 mL/0.2 g of DM when compared to T1, T2 and T4 which had 7.10, 9.10 and 10.40 mL/0.2 g of DM, respectively; being T1 the one with the lowest gas volume when compared to the rest of the treatments.

The importance of this variable is that the rate of gas production is proportional to microbial activity, but the proportionality decreases with incubation time, so it can be interpreted as the loss of fermentation rate efficiency with time (Lavrencic *et al.,* 1997).

Increased microbial activity predisposes to an increase in fibre digestibility, leading to high gas production during the initial hours of fermentation and thereby increasing voluntary feed intake (Fadel El-seed *et al.,* 2004).

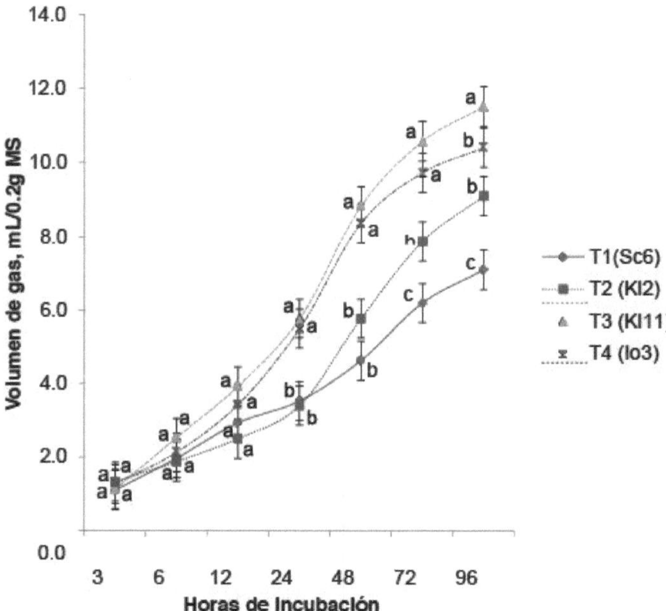

Graph 2. Means (± SE) of gas production during *in vitro* ruminal fermentation of T1, T2, T3 and T4 containing all treatments oat hay, corn silage, concentrate and their respective yeast strain. abc Means with different time letter are different (P<0.05) between treatments.

The degradation of a substrate starts with easily digestible fractions such as highly fermentable carbohydrates like starch. Therefore, the behaviour observed in this experiment suggests that the data obtained with higher digestibility of DM, NDF and FDA, as well as the lower LDA content in T2, T3 and T4 suggest that they may have favoured microbial colonisation, improving the digestibility of the diet.

***In vitro* fermentation parameters**

Table 16 shows the parameters of fermentation kinetics. Parameter **(A)** represents the time to turning point, being T2, T3 and T4 those who obtained the highest turning point (P<0.05) revealing 11.53, 11.6 and 10.42 h when compared to T1 which presented 8.05 h. This indicates that the strains of the treatments with higher turning point, perhaps favoured the microbial colonisation of the fibre causing an increase in the rate of fermentation and substrate, prolonging the turning point which reflects a higher volume of gas production.

Parameter **(B)** represents the mL of gas at the inflection point, where a significant difference was observed between treatments (P<0.05). T2, T3 and T4 presented the highest amount of gas at the inflexion point (2.21, 2.10 and 2.17 mL) when compared to T1 which revealed 1.68 mL. This parameter explains an increase in gas production, product of ruminal fermentation due to the improvement in the chemical composition of the diet, perhaps because strains KI2, KI11 and Io3 improved the microbial environment compared to Sc6, leading to an increase in fibre digestibility and therefore, an increase in the concentration of substrate available for the microorganisms (Cone *et al.,* 1998).

Table 16. Parameters of rumen DM digestibility of DM between treatments

Parameters	Treatment1				EE(±)
	T1	**T2**	**T3**	**T4**	
A	8.05^b	11.53^a	11.6^a	10.42^a	0.68
B	1.68^b	2.21^a	2.10^a	2.17^a	0.10
C	0.03^b	0.02^b	0.05^a	0.05^a	0.00

^{ab} Means with different row literals are different (P<0.05) between treatments.
A = Time to turning point.
B = mL of gas at the inflection point.
C = Maximum gas production rate (mL/h).
$_{1-T1}$ = Oat hay (HA) + corn silage (MS) + concentrate + *Saccharomyces cerevisiae* strain 6; T2 = HA + MS + concentrate + *Kluyveromyces lactis* strain 2; T3 = HA + MS + concentrate + *Kluyveromyces lactis* strain 11 and T4 = HA + MS + concentrate + *Issatchenkya orientalis* strain 3.

The biological function of this parameter suggests that strains KI2, KI11 and Io3 will produce 0.53, 0.42 and 0.49 mL more gas than Sc6 to reach the tipping point, indicating that they contain more substrate for the microorganisms.

The parameter **(C)**, represents the maximum gas production rate (mL/h), where it can be observed that T3 and T4 showed the highest (P<0.05) gas production rate, both with 0.05 mL/h while in T1 and T2 the estimates were 0.03 and 0.02 mL/h, respectively. This behaviour indicates that T3 and T4 have a greater ability to colonise the dietary particles than the rest of the treatments, resulting in a higher fibre digestibility and an increase in the volume of gas produced.

AGV Production and Profiling

Table 17 shows the VFA production and profiles between treatments, where it was observed that T2 presented higher (P<0.05) concentration of acetic, propionic and bitmic acid (37.30, 19.25 and 7.74 mmol/L) when compared to T1, T4 and T3. Likewise, T4 was higher (P<0.05) in propionic acid than T1 (16.59 and 12.52 mmol/L respectively). On the other hand, T1, T2 and T4 were superior (P<0.05) in butamic acid to T3 (7.45, 7.74, 6.73 and 4.51 mmol/L, respectively).

Allen and Mertens (1988) mentioned that the fermentation pattern in ruminants is influenced by the interaction between the diet, the microbial population and the animal itself. The addition of yeast cultures in ruminant diets has shown contradictory effects on rumen VFA concentration. On the other hand, it has been reported that the tendency to increase the molar ratio of acetate in the rumen of animals supplemented with yeast cultures, may be due to the fact that the rumen is a source of VFA in the rumen.

Table 17. Behaviour in VFA production and profile between treatments during *in vitro* fermentation.

Variables	Treatments^{VVI}				EE(±)
	T1	**T2**	**T3**	**T4**	
Acetic acid (mmol/L)	24.87^c	37.30^a	32.84^{ab}	29.91^{bc}	2.03
Propionic acid (mmol/L)	12.52^c	19.25^a	17.73^{ab}	16.59^b	0.89
Butamic acid (mmol/L)	7.45^a	7.74^a	4.51^b	6.73^a	0.52

^{abc} Means with different row literals are different (P<0.05) between treatments.
^{VI} T1 = Oat hay (HA) + corn silage (MS) + concentrate + *Saccharomyces cerevisiae* strain 6; T2 = HA + MS + concentrate + *Kluyveromyces lactis* strain 2; T3 = HA + MS + concentrate + *Kluyveromyces lactis* strain 11 and T4 = HA + MS + concentrate + *Issatchenkya orientalis* strain 3.

related to the improvement of the rumen environment, favouring the growth and activity of fibre degrading microorganisms (Chaucheyras *et al.,* 1995). In this investigation, T2 presented the highest concentration of acetic, propionic and butmco acids, this behaviour suggests that the Kl2 strain favoured the microbial environment by increasing the number of cellulolytic bacteria causing a higher DM and FDA digestibility, decreasing the LDA content as shown in Table 15. Dolezal *et al.* (2005) observed an increase in VFA production by increasing the dose of Sc yeast culture (strain SC-47) in the feed of lactating Holstein cows. Koul *et al.* (1998) reported an increase in total VFA in the rumen of buffalo calves fed 5 g/d of a Sc-containing yeast culture when compared to the control group (132.2 and 122.4 mmol/L). However, Harrison *et al.* (1988) found a decrease in the molar ratio of acetic acid and an increase in propionic acid concentration in the rumen fluid of Holstein cows supplemented with 114 g/d of a Sc-containing yeast culture.

Concentration of N-NH3, Lactic Acid and pH

Table 18 shows the N-NH3 and lactic acid concentrations and pH between treatments. The highest ($p < 0.05$) N-NH3 concentration was shown by T4 compared to T1, T2 and T3 (5.34, 3.06, 3.28 and 3.63 mM/mL, respectively). Similarly, the same treatment had higher ($p < 0.05$) concentration of lactic acid (0.69 mM/mL) compared to the rest of the treatments (0.44, 0.52 and 0.63 mM/mL for T1, T2 and T3). On the other hand, the pH in T1 was higher ($p < 0.05$) compared to T3 and T4 (7.00, 6.83 and 6.59, respectively).

Ammoniacal nitrogen (N-NH3). The digestion of proteins is related to their solubility in the rumen, where lower solubility decreases ammonia release. Moloney and Drennan (1994) reported that several sources of nitrogen contribute to ammonia production, such as non-protein nitrogen (NNP) from the diet, salivary nitrogen and possibly small amounts of urea that penetrate through the rumen epithelium.

In the present study the highest concentration of N-NH3 was observed at T4, this may be due to the fact that strain Io3 improved the microbial environment causing a higher activity of microorganisms, increasing DM and NDF digestibility and increasing the concentration of N-NH3 from fermentation. Fonty and Chaucheyras-Durand (2006) mentioned that the effect of yeast culture on N-NH3 concentration is highly variable, which depends on abiotic (diet) and biotic (rumen microorganisms) factors. In a study, Williams *et al.* (1991) reported that a decrease in rumen N-NH3 concentration in animals supplemented with yeast cultures may result in an increase in ammonia incorporation into microbial protein, which may lead to enhanced activity of rumen microorganisms, another possible reason may be a reduction in the activity of rumen proteolytic bacteria, as reported in the *in vitro* study by Chaucheyras-Durand *et al.* (2005). D^az (2011) reported similar behaviour of strain Io3 which increased N-NH3 concentration in yeast inocula during *in vitro* gas production.

Table 18. Behaviour of N-NH3 concentration, lactic acid and pH
between treatments during *in vitro* fermentation.

Variable	Treatments[1]				EE(±)
	T1	T2	T3	T4	
N-NHs (mM/mL)	3.06^b	3.28^b	3.63^b	5.34^a	0.39
Lactic acid (mM/mL)	0.44^c	0.52^{bc}	0.63^{ab}	0.69^a	0.06
pH	7.00^a	6.97^{ab}	6.83^b	6.59^c	0.05

[abc] Stockings with different verbatim in a row, are different (P<0.05) on treatments.

[1] T1 = Oat hay (HA) + corn silage (MS) + concentrate + *Saccharomyces cerevisiae* strain 6; T2 = HA + MS + concentrate + *Kluyveromyces lactis* strain 2; T3 = HA + MS + concentrate + *Kluyveromyces lactis* strain 11 and T4 = HA + MS + concentrate + *Issatchenkya orientalis* strain 3.

Lactic acid and pH. Martin and Nisbet (1992) mentioned that the incorporation of yeast cultures in ruminant diets helps to decrease lactate concentration in the rumen by stimulating lactate fermenting bacteria such as *Selenomonas ruminantium* (*Sel. ruminantium*) and *Megasphaera elsdenii* (*M. elsdenii*), preventing effects associated with lactic acidosis. Other studies have shown that *Aspergillus oryzae* (*A. oryzae*) and *Saccharomyces cerevisiae* (Sc) added to ruminant diets decrease lactate concentration by stimulating lactate-utilising bacteria such as *Sel. ruminantium* and *M. elsdenii* (Waldrip and Martin, 1993). In this experiment, the highest lactic acid concentration was found in T4, perhaps because strain Io3 does not stimulate lactic acid-utilising bacteria to a large extent, as indicated by several authors. D^az (2011) found similar behaviour of strain Io3 which increased the concentration of lactic acid in yeast inocula during *in vitro* gas production. On the other hand, ruminal pH reinforces the balance between buffering capacity and acidity of fermentation, even though an optimal pH cannot be defined in the rumen medium, microorganisms have a certain range in which they reproduce better and their metabolism is more efficient (Wales *et al.*, 2004). Low pH inhibits N-NH3 production *in vitro* causing a negative effect on methanogenic bacteria (*Methanobacterium bryantii*, *M. formicicum* and *Methanosarcina barkeri*) and protozoa in fibre degradation (Nagaraja and Titgemeyer, 2007).

In the present investigation, T1 had the highest pH when compared to T3 and T4, but was similar to T2, perhaps because the Sc6 strain optimised the microbial environment by maintaining pH neutrality, causing a decrease in the pH of the microbial environment.

lactic acid concentration. The reduction in the amount of lactic acid results in an increase in ruminal pH which favours the growth of cellulolytic bacteria, leading to an increase in fibre digestibility and VFA production (Chaucheyras-Durand and Fonty, 2001). However, under the conditions in which this research was carried out, the Sc6 strain did not favour VFA concentration and did not improve fibre digestibility when compared to strains Kl2, Kl11 and Io3. The addition of Sc to ruminant diets often results in an increase in the number of cellulolytic bacteria (*F. succinogenes, R. albus*), both *in vitro* and *in vivo* (Lila *et al.*, 2004). Chaucheyras *et al.* (1995) reported that Sc appears to stimulate lactate utilisation by *M. elsdenii and Sel. ruminantium*, resulting in increased propionate synthesis. Scharrer and Lutz (1990) mentioned that a decrease in ruminal pH is associated with an increased production of VFA and consequently reduces the concentration of N-NH3.

CONCLUSIONS AND RECOMMENDATIONS

The addition of yeast strains at T2 (Kl2), T3 (Kl11) and T4 (Io3) in the control diet of weaned calves favoured DM digestibility, with the last treatment showing an increase in NDF digestibility.

The addition of Kl (strain 2 and 11) to the diet increased the percentage digestibility of FDA, and together with Io3 showed a marked decrease in LDA content.

Strain KI11 showed the highest gas volume at 96 h of incubation *in vitro*. Likewise, together with Io3 they increased the maximum rate of gas production during ruminal fermentation. KI2 had the greatest increase in the concentration of acetic and propionic acids. On the other hand, KI2, Sc6 and Io3 increased butynic acid production.

On the other hand, Io3 showed an increase in the amount of N-NH3 and lactic acid, while Sc6 showed a clear increase in pH during *in vitro* fermentation.

Based on the results and the behaviour observed in this study of strains KI2, KI11 and Io3, further research is recommended, under different levels of inclusion, in order to achieve a better animal response.

LITERATURE CITED

Allen, M. S. and M. Mertens. 1988. Evaluating constraints on the fiber digestion by rumen microbes. J. Nutr. 118:261-270.

AOAC. 2000. Official methods of analysis. Vol. I. 16th ed. International Arlington. VA. U. S. A.

Bmck, H. and I. I. Turkmen. 2001. The effect of *Saccharomyces cereviciae* on *in vitro* rumen digestibilities of dry matter, organic matter and neutral detergent fibre of different forage: concentrate ratios in diets. Veteriner Fakultesi Dergisi, Uludag University. 20:29-37.

Broderick, G. A. and J. H. Kang. 1980. Automated simultaneous determination of ammonia and total amino acids in ruminal fluid and in vitro media. J. Dairy Sci. 63:64-75.

Brossard, L., F. Chaucheyras-Durand, B. Michalet-Doreau, and C. Martin. Martin. 2006. Dose effect of live yeasts on rumen microbial communities and fermentations during butyric latent acidosis in sheep: new type of interaction. Anim. Sci. 82:829-836.

Brotz, P. G. and D. M. Schaeffer. 1987. Simultaneous determination of lactic acid and volatile fatty acids in microbial fermentation extracts by gas-liquid chromatography. J. Microbiol. Methods. 6:139-144.

Callaway, E. S. and S. A. Martin. 1997. Effects of *Saccharomyces cerevisiae* culture on ruminal bacteria that utilize lactate and digest cellulose. J. Dairy Sci. 80:2035-2044.

Carro, M. D., P. Lebzein and K. Rohr. 1992. Influence of yeast culture on the *in vitro* fermentation of diets containing variable portions of concentrates. Anim. Feed Sci. Technol. 37:209

Chaucheyras, F., G. Fonty, G. Bertin and P. Gouet. 1995. *In vitro* H_2 utilization by a ruminal acetogenic bacterium cultivated alone or in association with an Archaea methanogen is stimulated by a probiotic strain of *Saccharomyces cerevisiae*. Applied Environ. Microbiol. 61:3466-3467.

Chaucheyras-Durand, F. and G. Fonty. 2001. Establishment of cellulolytic bacteria and development of fermentative activities in the rumen of gnotobiotically- reared lambs receiving the microbial additive *Saccharomyces cerevisiae CNCM I-1077*. Reprod. Nutr. Dev. 41:57-68.

Chaucheyras-Durand, F., N. D. Walker and A. Bach. 2008. Effects of active dry yeast on the rumen microbial ecosystem: Past, present and future.
Anim. Feed Sci. Technol. 145:5-26.

Chaucheyras-Durand, F., S. Masseclia and G. Fonty. 2005. Effect of the microbial feed additive *Saccharomyces cerevisiae CNCM I-1077* on protein and peptide degrading activities of rumen bacteria grown in vitro. Curr. Microbiol. 50:96-101.

Cone, J. W., A. H. Van Gelder and H. Valk. 1998. Prediction of nylon bag degradation characteristics of grass samples with the gas production technique. J. Sci. Food Agric. 77:421-426.

D^az, P. D. 2006. Production of microbial protein from waste apples added with urea and soybean paste. Master's thesis. Faculty of Animal Husbandry. Autonomous University of Chihuahua. Chihuahua. Chih. Mex.

D^az, P. D. 2011. Development of a yeast-based inoculum and its effect on in vitro fermentation kinetics in rations for high-producing Holstein cows. PhD thesis. Faculty of Animal Science and Ecology. Autonomous University of Chihuahua. Chihuahua. Chih. Mex.

Dolezal, P., J. Dolezal and J. Trinacty. 2005. The effect of *Saccharomyces cerevisiae* on ruminal fermentation in dairy cows. Czech J. Anim. Sci. 50:503-510.

Erasmus, L. J., P. H. Robinson, A. Ahmadi, R. Hinders and J. E. Garrett. 2005. Influence of prepartum and postpartum supplementation of a yeast culture and monensin, or both, on ruminal fermentation and performance of multiparous dairy cows. Anim. Feed Sci. Technol. 122:219-239.

Fadel El-seed, A. N. M. A., J. Sekine, H. E. M. Kamel and M. Hishinuma. 2004. Changes with time after feeding in ruminal pool sizes of cellular contents, crude protein, cellulose, hemicellulose and lignin. Indian J. Anim. Sci. 74:205-210.

Fonty, G. and F. Chaucheyras-Durand. 2006. Effects and modes of action of live yeast in the rumen. Biology. 61:741-750.

Goering, H. K. and P. J. Van Soest. 1970. Forage fibre analyses (apparatus, reagents, procedures and some applications). P. 379. In Agric. Handbook. USDA-ARS, Washington, DC. U. S. A.

Guedes, C. M., D. Goncalves, M. A. M. Rodrigues and A. Dias-da-Silva. 2008. Effects of a *Saccharomyces cerevisiae* yeast on ruminal fermentation and fiber degradation of maize silages in cows. Anim. Feed Sci. Technol. 145:27-40.

Harrison, G. A., R. W. Hemken, K. A. Dawson, R. J. Harmon and K. B. Barker. 1988. Influence of addition of yeast culture supplement to diets of lactating cows on ruminal fermentation and microbial populations. J. Dairy Sci. 71:2967-2975.

INAFED, 2008. National Institute for Federalism and Municipal Development. http://www.inafed.gob.mx/work/templates/enciclo/chihuahua/. Accessed September 10, 2012.

Karma, D. N., L. C. Chaudhary, S. R. Neeta Agarwal and N. N. Pathak. 2002. Growth performance, nutrient utilization, rumen fermentation and enzyme activities in calves fed *Saccharomyces cerevisiae* supplemented diet. Indian J. Anim. Sci. 72:472

Kholif, S. M. and M. M. Khorshed. 2006. Effect of yeast or selenized yeast supplementation to rations on the productive performance of lactating buffaloes. Egypt. J. Nutr. Feed. 9:193-205.

Koul, V., U. Kumar, V. K. Sareen and S. Singh. 1998. Mode of action of yeast culture (Yea-Sacc^{1026}) for stimulation of rumen fermentation in buffalo calves. J. Sci. Food Agric. 77:407-413.

Lavrencic, A., B. Stefanon and P. Susmel. 1997. An evaluation of the Gompertz model in degradability studies of forage chemical components. Animal Science. 64:423-431.

Lila, Z. A., N. Mohammed, T. Yasui, Y. Kurokawa, S. Kanda and H. Itabashi. 2004. Effects of a twin strain of *Saccharomyces cerevisiae* live cells on mixed ruminal microorganism fermentation *in vitro*. J. Anim. Sci. 82:1847-1854.

Martin, S. A. and D. J. Nisbet. 1992. Effect of direct fed microbials on rumen microbial fermentation. J. Dairy Sci. 75:1736.

Menke, K. H. and H. Steingass. 1988. Estimation of the energetic feed value obtained from chemical analysis and *in vitro* gas production using rumen fluid. Anim. Res. Dev. 28.7-55.

Moloney, A. P. and M. J. Drennan. 1994. The influence of the basal diet on the effects of yeast culture on ruminal fermentation and digestibility in steers. Anim. Feed Sci. Technol. 50:55-73.

Muro, R. A. 2007. Rumen degradation kinetics of three forage sources using in vitro digestibility by gas production. Doctoral dissertation. Faculty of Animal Science. Autonomous University of Chihuahua. Chihuahua. Chih., Mex.

Nagaraja, T. G. and E. C. Titgemeyer. 2007. Ruminal acidosis in beef cattle: the current microbiological and nutritional Outlook. J. Dairy Sci. 90:17-38.

Newbold, C. J., R. J. Wallace, X. B. Chen, and F. M. Mcintosh. 1995. Different strains of *Saccahromyces cerevisiae* differ in their effects on ruminal bacterial numbers *in vitro* and in sheep. J. Anim. Sci. 73:1811-1818.

Noguera, R. R., E. O. Saliba and R. M. Mauricio. 2004. Comparison of mathematical models for estimating degradation parameters obtained through the gas production technique. Livestock Research for Rural Development. Vol. 16, Art. No. 86. http://www.lrrd.org /lrrd16/11/nogu16086.htm. Accessed October 10, 2012.

Roger, V., G. Fonty, S. Komisarczuk-Bony and P. Gouet. 1990. Effects of physicochemical factor on the adhesion to cellulose avicel of the rumen bacteria. *Ruminococcus flavafaciens* and *Fibrobacter succinogenes*. Appliec Environ. Microbiol. 56:3081-3087.

SAS. 2004. SAS/STAT® 9.1 User's Guide. SAS. Institute Inc. United States of America.

Sauvant, D., S. Giger-Reverdin and P. Schmidely. 2004. Rumen Acidosis. Modeling Ruminant. Proceeding of the alltech's 20th annual symposium. Reimagining the feed industry, May 23-26,2004. Nottingham University Press. London. London. United Kingdom.

Scharrer, E. and T. Lutz. 1990. Effects of short chain fatty acids and K on absorption of Mg and other cations by the colon and caecum. Zeitschrift fur Ernahrungswissenchaft. 29:162-168.

Sewalt, V. J. H., W. G. Glasser, J. P. Fontenot and V. G. Allen. G. Allen. 1996. Lignin impact on fiber degradation. 1 quinone methide intermediates formed from lignin during in vitro fermentation of corn stover. J. Sci. Food Agric. 71:195-203.

Steel, D. R. G. and J. H. Torrie. 1997. Biostatistics. Principles and procedures. 2ª Ed. McGraw-Hill. Mexico.

Taylor, K. A. C. C. C. 1996. A simple colorimetric assay for muramic acid and lactic acid. Appl. Biochem. Biotechnol. 56:49-58.

Theodorou, M. K., B. A. Williams, M S. Dhanoa, A. B. McAllan and J. France. 1994. A simple gas production method using a pressure transducer to determine the fermentation kinetics of ruminant feeds. Anim. Feed Sci. Technol. 48:185-197.

Van Soest, P. J. 1994. Nutritional ecology of the ruminant. 2nd ed. Cornell University Press. Ithaca, New York. U. S. A.

Van Soest, P. J., J. B. Robertson, and B. A. Lewis. 1991. Methods for dietary fiber, neutral detergent fiber, and non-starch polysaccharides in relation to animal nutrition. J. Dairy. Sci. 74:3579-3583.

Waldrip, H. M. and S. A. Martin. 1993. Effects of an *Aspergillus oryzae fermentation* extract and other factors on lactate utilization by the ruminal bacterium *Megasphaera elsdenii*. J. Anim. Sci. 71:2770-2776.

Wales, W. J., E. S. Kolver, P. L. Thorne and A. R. Egan. 2004. Diurnal variation in ruminal pH on the digestibility on highly digestible perennial ryegrass during continuous culture fermentation. J. Dairy Sci. 87:1864-1871.

Whetten, R. and R. Sederoff. 1995. Lignin biosynthesis. The plan cell. 7:1001- 1013.

Williams, P. E. V., C. A. G. Tait, G. M. Innes, and C. J. Newbold. 1991. Effects of the inclusion of yeast culture (*Saccharomyces cerevisiae* plus growth medium) in the diet of cows on milk yield and forage degradation and fermentation patterns in the rumen of sheep and steers. J. Anim. Sci. 69:3016-3026.

Printed by Books on Demand GmbH, Norderstedt / Germany